Praise for
The Permaculture Earthworks Handbook

Understanding the hydrological cycle is key to restoring damaged landscapes and ecosystems. Even relatively small areas can benefit from well-chosen earthworks to invigorate the growth cycle. It is however essential to understand your landscape, geology and its hydrology before building your earthworks of choice or you may do more damage than good. Douglas Barnes has written a very enjoyable book with accessible technical information for different climates to enable the reader to understand this big subject. *The Permaculture Earthworks Handbook* is required reading for students of permaculture design, regenerative agriculture and earth restoration.

—Maddy Harland, Editor & Co-Founder,
Permaculture Magazine – practical solutions beyond sustainability

In *The Permaculture Earthworks Handbook*, Douglas Barnes has taken a complex subject and presented it in easy to understand language that anyone can follow. He leads you through the complete process of planning, design and execution for a variety of water harvesting systems including swales, ponds and dams. Permaculture strategies are applied throughout the process ensuring that you implement the best solution for your situation. A must-read for anyone undertaking an earthworks project.

—Robert Pavlis, author, *Building Natural Ponds* and *Garden Myths*

Creating earthworks to manage water resources is the crucial first step in permaculture land development. All other design work is built around the swales, drains, contour lines, pools and ponds that collect and direct rainfall. Drawing on deep experience, with detailed exploration of the cycles of rainfall, the flow of water on the land, soil types, landform, and the tools and techniques of earthworks, Douglas Barnes provides an essential guide doing it right the first time.

—Darrell Frey, author, *Bioshelter Market Garden: A Permaculture Farm*
and Co-author, *Food Forest Handbook*

We came from water. We live on a water world. We are water. Climate change is weirding our weather but, overall, it will make more rain, not less. That is how the sky cleanses itself of carbon. How that rain falls, what it does when it hits ground, and whether it will be there when you need it all depend on many moving parts, but Douglas Barnes has given us the tool kit we have long needed to make the most of this. *The Permaculture Earthworks Handbook* transforms dilemma into opportunity. This needs to be a standard reference on the desk of every landscape designer, forester, agronomist and master planner.

—Albert Bates, author, *The Post-Petroleum Survival Guide and Cookbook,*
The Biochar Solution, and *The Paris Agreement*

In *The Permaculture Earthworks Handbook*, author Douglas Barnes uses clear, accessible language to explain everything from ponds, dams and swales to contour bunds, microcatchments, and hugelswales, from bench terracing and river training to pattern planting, spate irrigation and many other sophisticated techniques for harvesting water. And best of all, the book's strategies rest firmly on fundamental permaculture principles such as zone and sector planning, stacking, functional connectivity, efficiency and flow.

This remarkably thorough and comprehensive book will be of great value to all who are seriously interested in shaping the ground to get the maximum value from the water that is available to them, while also taking excellent care of their land.

—Sue Reed, Landscape Architect, author, *Energy-Wise Landscape Design*
Co-author, *Climate-Wise Landscaping*

Douglas Barnes offers an abundance of detail on harnessing the power of water for your homestead. After reading this book, you'll be ready to move some serious dirt.

—Rebecca Martin, Managing Editor for *Mother Earth News*

Water is the key element wherever we choose to live. Having travelled widely with indigenous people in both wet and dry climates I know how much care we need to spend on getting water if we are to break with the unsustainable option of plumbing into the grid. Douglas Barnes has written the key book on harvesting water for permaculture living. Clear, lively and extremely well written — it's destined to be a classic.

—Robert Twigger, bestselling author, *Micromastery* and *Angry White Pyjamas*

THE
PERMACULTURE EARTHWORKS
HANDBOOK

THE PERMACULTURE EARTHWORKS HANDBOOK

HOW TO DESIGN AND BUILD SWALES, DAMS, PONDS, AND OTHER WATER HARVESTING SYSTEMS

DOUGLAS BARNES

new society
PUBLISHERS

Cover design by Diane McIntosh.
Cover photo © David Spicer (Location "Eagles Deep"). Background map image: © iStock.
P 1: © Stramyk Igor; p. 14, 20: © Marina; p. 25: © EddieCloud; p. 56: © arenaphotouk;
p. 57: © nd700; p 58: © dihydrogen; p. 59: © LovePhy; p. 102: © Vevchic/Adobe Stock.

Printed in Canada. Third printing May 2023.

This book is intended to be educational and informative. It is not intended to serve as a guide. The author and publisher disclaim all responsibility for any liability, loss or risk that may be associated with the application of any of the contents of this book.

Inquiries regarding requests to reprint all or part of *The Permaculture Earthworks Handbook* should be addressed to New Society Publishers at the address below.
To order directly from the publishers, please call toll-free (North America) 1-800-567-6772, or order online at www.newsociety.com

Any other inquiries can be directed by mail to:

New Society Publishers
P.O. Box 189, Gabriola Island, BC V0R 1X0, Canada
(250) 247-9737

LIBRARY AND ARCHIVES CANADA CATALOGUING IN PUBLICATION

Barnes, Douglas, 1969–, author
The permaculture earthworks handbook : how to design
and build swales, dams, ponds, and other water harvesting
systems / Douglas Barnes.

Includes bibliographical references and index.
Issued in print and electronic formats.
ISBN 978-0-86571-844-9 (softcover).—ISBN 978-1-55092-639-2 (PDF).—
ISBN 978-1-77142-234-5 (EPUB)

1. Water—Storage—Handbooks, manuals, etc. 2. Water harvesting—
Handbooks, manuals, etc. 3. Permaculture—Handbooks, manuals, etc.
I. Title.

TD430.B37 2017   628.1'3   C2017-904852-X
                                                      C2017-904853-8

Funded by the Government of Canada · Financé par le gouvernement du Canada | Canada

New Society Publishers' mission is to publish books that contribute in fundamental ways to building an ecologically sustainable and just society, and to do so with the least possible impact on the environment, in a manner that models this vision.

new society PUBLISHERS

Certified B Corporation

FSC MIX Paper from responsible sources FSC® C016245

To my father.
And to everyone who sees solutions
and has the passion and tenacity to act.

Contents

Introduction

In every society throughout history, water has always been the vital ingredient that makes life possible. Without adequate water supplies, survival is simply not possible. This importance is reflected in cultures around the globe. With growing populations and climate change, water is becoming more important than ever, and not just in parts of the world that experience water scarcity. We already live in a time of water refugees as large numbers of people are forced to migrate due to insufficient water supplies. There are also increasing reports of violent conflict around water rights. Tensions can and do rise across and inside borders surrounding access to water. Conflict around water will rise as strain on available supplies increases.

For permaculture design, water is the starting point around which a site's plan will unfurl. Water-harvesting earthworks are an aspect of site design that can be the difference between a site that performs poorly and one that thrives. On damaged sites without adequate water, the right earthworks can have unbelievable results. Few things are as exciting as watching a dying site come back to life as soon as the first rains hit the project site.

It is true that humans have done tremendous damage to the Earth. The destruction has been centuries in the making, and tremendous energy has been put into the destruction. Yet herein lies great hope for the future. It has taken colossal amounts of effort, energy, and time to degrade the planet to its present state. With a nearly infinitesimal portion of the energy and time needed to create the destruction, we can heal the damage that has been done. There is no law of nature that humans must be a destructive force on the planet. In fact, through our actions, we can greatly foster life while meeting our own needs. The simple beaver is able to create diverse habitats that become havens for all manner of plants and animals. While beavers are impressive, human beings have the power of creativity and a curiosity that provides us with greater and greater understanding of how nature works. We already have the knowledge to have a greater positive effect on the environment, and our knowledge is growing every year. This book is a look at how you can lay the groundwork to make this possible.

The book is divided into nine chapters and has six appendices to explain how to handle the necessary calculations needed for the earthworks described in the book. Chapter 1 looks at the state of water in the world today and the challenges we face for the future. Chapter 2 looks at one of the most remarkable pieces of water engineering in human history, the ancient Nabatean city of Petra. Petra provides both excellent models for water capture and storage, and a cautionary tale around the inherent dangers in earthworks. Chapter 3 takes a close look at the interactions between soil and water. It also provides some conceptual models for looking at landscapes. This knowledge is very helpful in designing earthworks, as it gives an understanding of water's behavior on the land. Chapter 4 looks at the permaculture design process and its strategies for dealing with the design of complex systems, such as water-harvesting systems. It provides a mental template by which you can look at and design a landscape. Chapter 5 covers specific site aspects that you will need to address in the course of designing and implementing a project on any site, including climate. It also covers the typical tools and machinery involved in design and implementation. Chapter 6 looks at specific techniques for water storage and where those approaches should and should not be used. Chapter 7 covers water interception techniques and cautions surrounding them. Chapter 8 looks at integrating the techniques from Chapters 6 and 7 with the strategies used in permaculture design. Chapter 9 looks at the risk involved when employing water-harvesting earthworks. This chapter will help you to identify when earthworks are a hazard on a site so that you can better know when not to use them. There are also six appendices at the end, providing the equations you will need for designing and costing water-harvesting earthworks.

The book is written so that each chapter lays the foundation for later chapters. As such, it is recommended that you read through the whole book in order, rather than diving straight into the techniques in Chapters 6 and 7. This will give you a better understanding of the subject and will help to prevent disastrous errors that can occur from doing the wrong thing in the wrong place.

The State of Water

The Colorado

On March 25, 2014, near the Sea of Cortez in the Sonoran Desert, a jovial crowd gathered in a dry, sandy riverbed flanked by cottonwood and willows on either side. They were there to witness a rare event. The occasion that attracted so much attention was a trickle of water moving along the dry riverbed at the speed of a lazy stroll. Two days earlier, the Morelos Dam had slowly opened the gates to the Colorado River.

This artificial mimicking of the natural spring flows that used to occur was a result of Minute 319 of the International Boundary and Water Commission. On November 20, 2012, both the United States and Mexico agreed to the goal of working toward the restoration of the Colorado River. This was the first time that a water allocation on an international river was made strictly for the environment.

Two months after the release from the dam, the flow of water, dubbed "the pulse," finally reached the Sea of Cortez on May 15. Three days later the Morelos Dam was once again closed, and the pulse ended. While the next four years were to see additional base flows released, these smaller allocations were, in total, less than the pulse flow of 2014. The ongoing base flows have helped to rejuvenate the lower Colorado, and in July of 2016 a sea lion was spotted in the upper estuary for the first time.

What made the pulse so special? Why was an international agreement necessary to restore a fraction of the water that had once fed a thriving, 3,000-square-mile delta? Since the completion of the Hoover Dam in 1936, ten dams have been built along the main stem of the Colorado River—this in addition to the three dams that preceded the Hoover Dam. Add to this the thirty-one major dams along the tributaries of the Colorado, as well as the irrigation channels built into the river system, and it becomes easy to see how the Colorado's flow never reached the sea.

Over the course of the 20th century, the river had come to be claimed for a human population that would grow to 30 million people. It became the source of power generation, irrigation, and municipal water supplies, but the success of these engineering projects came at the expense of the natural environment that ultimately supports those same people.

The Aral

Though the Colorado story has a glimmer of hope to it, a similar story on the other side of the world is an ongoing crisis on a far greater scale. In the 1950s, the Soviet Union redirected the Amu Darya and Syr Darya rivers in order to support desert agriculture in the area around the Aral Sea. The Aral Sea itself was dependent on those rivers to maintain its volume. Without the flow from the rivers, the Sea started to evaporate, leaving behind negative health effects and a ruined economy for tens of millions in the region. Infant mortality rose to a staggering 1 in 10; tuberculosis deaths rose 21 times higher; cancer saw a 10-fold increase; kidney disease rose 15 times higher; and gastritis deaths went up by 15 percent. To add insult to injury, up to 75 percent of the redirected water was wasted.

The water problems didn't end with an evaporating sea. The loss of volume of the sea had a corresponding loss of groundwater levels. This loss of groundwater, in turn, led to increased salinization of the soils of the region. This hindered local plant growth, contributing to erosion, which in turn led to a dependence on fertilizers for agriculture.

Dust storms are now a regular occurrence, with the salt content in the dust being as high as 90 percent, increasing respiratory illness. This salt can be carried a long distance, having harmful effects on agriculture far from the sea itself.

Through evaporation, the Aral split into the North and South Aral Seas in 1990. At that point, the Royal Geographical Society called the Aral Sea "the world's worst disaster." In an attempt to prevent the North Aral Sea from draining, a sand dam was built in the mid-1990s, though it had failed by the end of the decade. With funding from the World Bank, a new dam was completed in 2005, and since that time, the North Aral Sea has risen over 10 meters (32.8 feet), which has led to a revitalization of the fishing industry.

Talupula

Water crises also strike many communities on a local scale. Such is the case for Talupula, a remote village in Andra Pradesh, India. Once a dry tropical region, it has been growing increasingly arid over the decades, and the life-giving monsoons have become less reliable. Overgrazing and the harvesting of forests for fuel has denuded most of the landscape. During the dry season, the region has the look of a desert. This loss of vegetation has reduced the land's capacity to capture and store water. This, in turn has reduced the rate of groundwater recharge. The town relies on an aquifer over 1,000 feet deep; and the rate of abstraction is lowering the water level year by year. The biotic pressures on the landscape have diminished the recharge rate of the aquifer. While redirecting and damming river flows are not the culprits here, anthropogenic changes to the watershed are.

To compound problems, fluorite, fluorapatite, and other minerals in the rock leave the water heavily fluoridated, making fluorosis a health concern. From a health standpoint, the water is considered unsuitable for drinking, yet it is the only current option for the town's supply. High fluoride levels can also affect livestock reproduction and plant germination and growth. The high evaporation rate and low rates of recharge are suspected of compounding the fluoride issue.

Like every place in the world with a crisis looming over it, life chugs on, albeit with a sense of hopelessness in many of the residents. The environmental changes are progressing at a rate that even the young can perceive. And yet, as we will see in Chapter 7, Talupula offers an exciting ray of hope. As part of a cooperative project with the Green Tree Foundation of AP, India, we were able to turn a barren hillside into a mango orchard for under $1,000, using very simple earthworks.

Worldwide

Globally, humans use 4,000 km^3 of water each year. Of this, 70 percent is used for agriculture, 20 percent for industrial purposes, and 10 percent for domestic use. It should be noted that a portion of the agricultural usage is now tied into energy production with biofuels.

With population increasing and climate change growing more severe, the World Bank estimates that Central Africa and the Middle East will lose 6 percent of their GDP to water scarcity. For with a 2°C increase in global average temperature, the percentage of the global population affected by absolute water scarcity (meaning that individuals have less than 500 m^3 water per year) is predicted to increase by 5 to 20 percent, and the population experiencing water scarcity (less than 1,000 m^3 per year per individual) is predicted to increase by between 40 and 100 percent, depending on population rates and warming rates.

To meet the needs of decreasing water supplies, groundwater is being drawn on at increasing rates. Currently, 48 percent of agriculture globally relies on declining supplies of groundwater for irrigation. Over the coming decades, the declines in groundwater supplies will severely limit agriculture in many regions. In addition to human depletion of groundwater, climate change is also threatening supplies.

The rate of abstraction globally has increased threefold over the past 50 years and is increasing at 1 to 2 percent per year. It is estimated that abstraction will increase by a further 55 percent by 2050. Currently, 26 percent of the global water supply is provided by groundwater, and almost half of the water extracted is used to meet the need for potable water. This rate of withdrawal currently amounts to 8 percent of the mean global aggregate of groundwater recharge, but reliance on groundwater abstraction is greatest in regions of water scarcity. It is in these regions that recharge is often less than abstraction. Over-abstraction puts tremendous pressure on agriculture, threatening the food security of already impoverished nations. The result is often environmental refugees fleeing land that can no longer support them. For instance, roughly two thirds of India's irrigation needs are met by groundwater, and wells are being abstracted at a rate greater than they can recharge.

Degradation of groundwater is also an increasing problem. Over-abstraction in some coastal areas has allowed saltwater to backfill, contaminating the groundwater. Here, too, climate change is expected to exacerbate the problem as sea levels rise, putting more water systems at risk.

Tracking groundwater is difficult, and we have only a rough picture of what reserves are in place and just how much they are declining. In 2002, NASA started tracking groundwater levels with its GRACE mission (Gravity Recovery and Climate Experiment). From the data it has collected, we know that one third of the Earth's large groundwater basins are being depleted at an alarming rate.

Spread of deserts

Two billion people live in drylands globally, most of them below the poverty line. Accounting for 41.3 percent of all land and 44 percent of cultivated land, and containing 50 percent of the world's livestock, these areas are increasingly coming under threat of desertification. Deforestation, farming practices, mining, and climate change are increasing the spread of deserts across dryland areas.

As drylands further degrade, they are expected to lower the global production of food by 12 percent over the next 25 years, raising food prices by some 30 percent, leaving nearly a billion people hungry. We lose 23 hectares a minute (over 12 million hectares a year) to desertification.

The deforestation and degradation of Talupula has been a process many decades in the making. It took massive engineering projects to choke off the Aral Sea and to use up the Colorado River before the river could reach the ocean. The Earth's water problems have been centuries in the making. To do the damage we have done has taken tremendous energy and billions of labor hours and machine hours. To put it succinctly, it has taken a lot of work to muck up the Earth to the extent we have.

One obvious and important lesson to be gleaned from the Aral Sea, Colorado River, and even Talupula is that what happens upstream affects what happens downstream. We also see that large-scale engineering projects can lead to large-scale problems. Both the drying of the Aral Sea and the reduction of the Colorado River have come about through inappropriate approaches to irrigation. These two stark examples are played out on a less dramatic scale throughout much of the world today.

These three examples also show us a ray of hope. The explosion of life in the Colorado Delta, the return of fisheries to the North Aral Sea, and the results of Talupula show us just how quickly systems can be rejuvenated when we add water.

Water is vital for all life on the planet. No water, no life. We need water to survive. We need water to produce the food we eat. We also use water in cooking and cleaning processes, sanitation, and to manufacture the items we need in our daily lives.

War and conflict

With growing water scarcity comes an increase in conflict over water. This conflict takes place not only between nations but also within nations as well. In January 2014, for instance, a small dam in Kyrgyzstan was targeted with mortar rounds by Tajik security forces as part of a conflict over water- and pasture-access rights. Water supply systems are also used as targets in terrorist attacks. Such was the case in an attack on Afghan schoolgirls in April 2012, when their school's water supply was poisoned out of opposition to the education of girls. Class conflict within a nation, too, has arisen around water rights. Growing water scarcity in the small village of Rasooh, in the state of Jammu and Kashmir, India, for instance, has seen assaults on lower caste Dalit women attempting to access village well water. Police were then needed to guard the well. The well was later damaged by a group of members of the upper caste to show their disapproval of the lower caste accessing the well.

These are but three small examples of a growing number of conflicts involving water. Africa, Asia, Australia, Europe, and North and South America all have recent records of violence or attempted violence involving water. While there are major international conflicts involving water, such as the ongoing Syrian Civil War, there is a growing number of lesser national and international conflicts around water. Increased water scarcity, together with an increasing population and the spread of deserts, makes the risk of increased future conflict more and more likely. Securing sustainable water supplies for populations at risk will increasingly become a prerequisite for peace.

Where there is hope

While the sheer numbers of people and the advent of machinery have accelerated the process, the degradation of drylands has been millennia in the making. The catastrophes of the Colorado River and the Aral Sea both required massive engineering projects. Aquifer depletion has required electric and fossil-fuel–powered pumps to withdraw water at rates greater than recharging. Damaging the environment is not that easy to do. It takes tremendous concentrated effort to really muck things up. Yes, humankind has made a real mess of much of the globe, but not without expending trillions of labor hours and quadrillions of kilocalories to do so. Simply put, breaking the planet is hard work.

The really good news is that by working in cooperation with nature, we can undo most of the damage we have done with a fraction of the time and energy it took to cause the damage in the first place. I never cease to be amazed when, time and time again, degraded systems start to turn around immediately after the first rainfall on the repair site. Earth repair is one of those rare cases in which fixing the mess is easier than making it in the first place. This book is about the first steps toward making that repair.

Just add water!

A successful ecological system requires water. There is no way around this. Without water, there is no life. In the Earth's drylands, the importance of water is no mystery. In the wetter regions, however, it is often overlooked. In Ontario, Canada, where I grew up and currently live, water is often seen as a nuisance in the spring— something that can interfere with planting. Yet drought is an all-too-common occurrence in the late summer. Many Ontario farmers just surrender to the mercy of the weather and hope for adequate rainfall during the growing season. Whether you are living in drylands, a humid temperate climate, or a rainforest, water-harvesting earthworks have the potential to improve agriculture and local ecosystems.

Humans are a part of the biosphere. Through both our history and our prehistory, we have made some real ecological blunders. Yet we have the potential not only to meet our own needs but also to foster biodiversity and ecosystem health while meeting those needs. We can be good stewards of the land instead of the biological equivalent of rowdy hooligans.

References

Arveti, N., M.R. Sarma, J.A. Aikenhead-Peterson, and K. Sunil. "Fluoride incidence in groundwater: a case study from Talupula, Andhra Pradesh, India." *Environmental Monitoring and Assessment* 2011-01 doi: 10.1007/s10661-010-1345-3.

Barstow, Lynne. "Raise the River. The Colorado River Delta Pulse Flow: 1 year later." 2015-05-20. raisetheriver.org/the-colorado-river-delta-pulse-flow-1-year-later/.

British Broadcasting Corporation (BBC). "On This Day 1050-2005: 22 October. 1990: Aral Sea is 'world's worst disaster'." 2008. news.bbc.co.uk/onthisday/hi/dates/stories/october/22/newsid_3756000/3756134.stm.

Columbia University. "The Aral Sea Crisis." 2008. columbia.edu/~tmt2120/introduction.htm.

EurasiaNet. "Kyrgyzstan-Tajikistan: What's Next After Border Shootout?" 2014-01-13. eurasianet.org/node/67934.

Gleick, Peter H., et al. *The World's Water, Volume 8: The Biennial Report of Freshwater Resources.* Washington DC: Island Press, 2014.

International Bank for Reconstruction and Development/The World Bank. "High and Dry: Climate Change, Water and the Economy." 2016. openknowledge.worldbank.org/bitstream/handle/10986/23665/K8517.pdf.

Jet Propulsion Laboratory. "Study: Third of Big Groundwater Basins in Distress." 2015-06-16. jpl.nasa.gov/news/news.php?feature=4626.

Kendy, Eloise. "The Nature Conservancy. Colorado River: Six Months After the Pulse Flow." nature.org/ourinitiatives/regions/northamerica/areas/coloradoriver/colorado-river-six-months-after-the-pulse-flow.xml.

National Aeronautics and Space Administration (NASA). "NASA GRACE Data Hit Big Apple on World Water Day." 2012-03-22. nasa.gov/mission_pages/Grace/news/grace20120322.html#.WBkmCeErIWp.

OECD. "OECD Environmental Outlook to 2050: The Consequences of Inaction—Key Facts and Figures." 2012. oecd.org/env/indicators-modelling-outlooks/oecdenvironmentaloutlookto2050theconsequencesofinaction_keyfactsandfigures

Ortloff, Charles R. *Water Engineering in the Ancient World: Archaeological and Climate Perspectives on Societies of Ancient South America, the Middle East, and South-East Asia.* Oxford: Oxford University Press, 2009.

Pacific Institute. "Water Conflict Chronology List." worldwater.org/conflict/list/.

Schewe, Jacob, Jens Heinke, Pavel Kabat, et al. "Multimode assessment of water scarcity under climate change." *Proceedings of the National Academy of Sciences of the United States of America*, vol. 111, no.9 (2014-03-04) doi: 10.1073/pnas.1222460110.

Sharma, Ashutosh. Thomson Reuters Foundation. "Water scarcity heightens caste tensions in India." 2014-01-30. news.trust.org//item/20140129120744-te896/?source=hptop/.

United Nations Convention to Combat Desertification (UNCCD). "Desertification." unccd.int/Lists/SiteDocumentLibrary/Publications/Desertification-EN.pdf.

United Nations Convention to Combat Desertification (UNCCD). "Desertification Land Degradation & Drought (dldd)—Some Global Facts & Figures." unccd.int/Lists/Site DocumentLibrary/WDCD/DLDD%20Facts.pdf.

United Nations. "World Day to Combat Desertification." un.org/en/events/desert ificationday/background.shtml.

UN-Water. "A Post-2015 Global Goal for Water: Synthesis of key findings and recommendations from UN-Water." 2014. unwater.org/app/uploads/2017/05/UN-Water _paper_on_a_Post-2015_Global_Goal_for_Water.pdf.

van der Gun, Jac. *Groundwater and Global Change: Trends, Opportunities and Challenges*. 2012. unesco.org/fileadmin/MULTIMEDIA/HQ/SC/pdf/Groundwater%20and %20Global%20Change.pdf.

Warwick, Joby. "A River Once Ran Through It: The Colorado River Delta was an oasis for wildlife and people until the water stopped flowing." 2002-01-02. nwf.org/News -and-Magazines/National-Wildlife/News-and-Views/Archives/2002/A-River-Once -Ran-Through-It.aspx.

Weiss, Margo. "Phys.org. Research indicates 60-year decline in groundwater levels across US." 2014-03-31. phys.org/news/2014-03-year-decline-groundwater.html.

A Look at the Past

One of the key requirements for the existence of a civilization is water. If water is not present in sufficient amounts, civilization is simply not a possibility. Every civilization, from tribal stone-age nations to the most technologically advanced nation of a given age, has had its own water infrastructure. It is usually at the more extreme ends, however, that we see the most remarkable water systems, as well as the most creative. We will look at the approaches and outcomes of the ancient Nabatean city of Petra.

Petra

The archeological record for the Nabatean city of Petra spans from 300 BC to the 7th century AD. The architecture of the city reflects the influence Petra had as an important trade stop on the Silk Road. Techniques and styles were borrowed from the region, including Greek, Roman, and Syrophoenician. In the development of its water capture and distribution system, Petra was a world leader, and there was little that other cultures had to offer in terms of improvements.

Located in modern-day Jordan, between the Dead Sea and the Gulf of Aqaba at Jebel al-Madhbah Mountain in the Sharah mountain range, Petra receives 177 millimeters of rain per year. While the climate has certainly changed to some degree over the centuries, the record of the region's aridity suggests that things are not too different now from in ancient times. Furthermore, pollen analysis shows that the destruction of forests in the Ghab Valley in Syria had already started as far back as 9,000 years ago. The cedar forests of Lebanon began to be cleared 7,700 years ago. To put this into context, these deforestations began 4,000 and 3,000 years before the time of the *Epic of Gilgamesh*, respectively. It is worth noting that the prime culprit for the great flood mentioned in *Gilgamesh* is the deforestation of headwaters.

For the crimes of killing the forest protector Humbaba and cutting the sacred trees, the god Enlil punishes the main character, Gilgamesh. Forest hydrology is covered here in some detail in Chapter 7, but the effect of deforestation is to both increase runoff and decrease total rainfall. The ancient deforestation would help to explain both the aridity of the region and the flood legends. The resulting transformation of the climate can be viewed as punishment for the cutting of the regions forests.

Given the aridity of the region, how was it that Petra was able to support between 20,000 and 30,000 people? Where did the water come from? How was it made available to the people? And what happened to the city to make people abandon it? The answers to these mysteries can lend clues as to how we can make better use of our own water resources.

Petra had two options for water, both of which were used: harness springs and harvest rainwater. The main water supply was the spring at Ain Musa, seven kilometers outside of the city. The pipe system to deliver this water was a marvel of hydrological engineering for the time. However, it is the network of water-harvesting systems that is of interest for the purposes of this book.

In addition to springs, Petra employed dams, cisterns, and water-distribution tanks to meet the city's water needs. The Wadi Musa River, originally flowing through the iconic canyon at Petra, was diverted by a stone embankment, leading water through a flood bypass tunnel whereby the water would enter the city from the northeast. Dozens of stone-wall dams were used throughout the city and surrounding hills to control flood water, capture runoff, and allow silt to drop out of water to improve water quality. Additionally, more than two hundred surface and underground cisterns were used to capture runoff to provide more water for the city. Leakage from dams helped to recharge ground water and raise the water table. This would have benefited spring flow as well as help to sustain water levels in underground cisterns.

Petra's diverse approach to water supply provided redundancy, which ensured that water was available during times in which an individual water supply element or group of elements needed maintenance, repair, or modification. It also had the capacity to provide several months' worth of water in reserve. The system worked well enough that at the city's peak population of 30,000 people there was a per capita water budget of 40 liters (10.6 US gallons) per person. The minimum requirement for survival in the region's hot desert climate is 3 liters (0.8 US gallons) per day. Petra had enough water for public use in the form of watering troughs, baths, fountains, and water gardens.

The system also had a 43-meter-long, 23-meter-wide, and 2.5-meter-deep pool (141 feet by 75.5 feet by 8 feet), with a capacity of over 2 million liters (528,000 US gallons)—a conspicuous display, signifying the city's regional importance. From the pool, the flow took the water on to a castellum, where it flowed in several directions to gardens, baths, workshops, and finally to a tunnel to Wadi Musa, where it flowed off into Wadi Siyagh. In this way, the water was put to multiple uses before flowing off site.

Petra has what was arguably one of the most brilliant water systems of its time. So what happened? Why was the city abandoned? Petra's demise appears to have come from two fronts: economic and environmental.

On the economic front, Petra's position as a Silk Road hub was eroded by sea travel, shifting regional power, and political changes. Frankincense and myrrh were the major goods moved through Petra, with processing of products taking place in the city before the goods were moved out for trade. In 106 AD, Roman emperor Trajan declared Bosra, in modern-day southern Syria, the capital of the Province of Arabia, moving the center of trade and power northward. This aided the city of Palmyra, also located in modern-day Syria, allowing it to supplant Petra's trade position due to its establishment of a more securely protected trade route.

For the earthworks designer, the environmental factors of Petra's decline are more instructive. In May of 363 AD, Petra was struck by a devastating earthquake. The quake damaged buildings in the city but, more importantly, would have damaged the water systems—specifically the flood-control systems that were integrated into the dams used for water harvesting.

At some point between the 363 earthquake and the 5th century, a massive rain event along with the failure of one or more dams in the city, released a great torrent of water through the city's core. The infrastructure that made life in the city possible ironically delivered the death blow that led to the eventual abandonment of Petra. Examination of the alluvial deposits in the city, as well as the pattern of swept-away paving stones, suggests that the collapse of the dam saw a surge of water from 4 to over 6 meters deep (13 to 20+ feet), moving at a maximum speed of over 3 meters per second (6.7 miles per hour), that plowed through the city, leaving destruction in its wake.

Petra would have been faced with more than just tending to the injured and rebuilding the damage to the city. It would have been left with a destroyed water supply and flood-control system. With flood diversion and control crippled, the city was left at risk of future flash floods causing more death and destruction in the city.

There are important lessons in the amazing ingenuity of Petra's water-harvesting systems, as well as a cautionary tale of the potential for disaster in the failure of those systems. The Nabatean approach of using springs while also capturing all available runoff was a novel approach that remains instructive to this day. Out of a desert, they were able to support a peak population of 30,000 people, providing water in abundance. From a safety perspective, however, the placement of dams in the wadis, upstream from buildings also built in the wadis, was a disaster that was inevitable.

References

Bedal, Leigh-Ann, and Bruce E. Rutledge. *The Petra pool-complex: A Hellenistic paradeisos in the Nabataean capital (results from the Petra "Lower Market" survey and excavation, 1998)*. PhD diss., University of Pennsylvania, 2000.

Bryce, Trevor. *Ancient Syria: A Three Thousand Year History*. Oxford: Oxford University Press, 2014.

Cummins, Dennis P. *The Role of Water in the Rise, Prominence, and Decline of Nabatean Petra*. Master's diss., The City University of New York, 2014.

Ghana, A. "Case Study: Trends and Early Prediction of Rainfall in Jordan." *American Journal of Climate Change* (2013) doi: 10.4236/ajcc.2013.23021.

Ortloff, Charles R. *Water Engineering in the Ancient World: Archaeological and Climate Perspectives on Societies of Ancient South America, the Middle East, and South-East Asia*. Oxford: Oxford University Press, 2009.

Paradise, Tom. "The Great Flood of Petra: Evidence for a 4th–5th AD Century Catastrophic Flood." *Annual of the Jordan Department of Antiquities* (2012).

Yasuda, Y., H. Kitagawa, and T. Nakagawa. "The earliest record of major anthropogenic deforestation in the Ghab Valley, northwest Syria: A palynological study." *Quaternary International* (2000-11) doi: 10.1016/S1040-6182(00)00069-0.

How Water Moves in the Environment

3

Fresh water is the primary concern on any site. If it is available in ample supply, it might appear not to be important, but it is the vital ingredient that makes all the living systems on the site possible. Every site can benefit from a well-designed water-management system. The prerequisite to designing an appropriate, well-functioning water-harvesting scheme is understanding the behavior of water in soils.

The bulk of the planet's water is salt water, carrying dissolved minerals that only specially evolved systems can take advantage of. Only 6 percent of the Earth's total water supply is fresh water, the overwhelming majority of which is bound up in groundwater and ice. The remaining amount—less than 1 percent—is what we have to work with. To be more precise, our main concern is making the best use of the atmospheric water that becomes available to us through precipitation. Table 3.1 shows where the world's water supply is, as well as the time that the water is resident in that sector of the hydrological cycle.

The hydrological cycle

The hydrological cycle is the global process of the circulation of water moving through the atmosphere to the land and on to the oceans. The ocean, the atmosphere, groundwater, and lakes can be thought of as reservoirs through which water cycles. Solar radiation powers the cycle by transforming water to vapor. From the atmosphere it precipitates onto the land, where it is governed by gravity, capillary action, and biological processes (see Table 3.1).

The total hydrological cycle is delineated on land into catchments, which are distinguished by drainage back to the sea. The area from which fallen precipitation drains into a basin via a single stream or river is the catchment. This is gravity driven, meaning that the topography defines the boundaries of the catchment.

Table 3.1. World water allocation and residency time.

Location	Percentage of total water supply	Average time in residence
Oceans and seas	94%	4,000 years
Groundwater	~4%	2 weeks to 10,000 years
Icecaps and glaciers	~2%	10 to 1,000 years
Lakes and reservoirs	Less than 0.01%	10 years
Soil moisture	Less than 0.01%	2 weeks to 1 year
Atmospheric water	Less than 0.01%	10 days
Rivers	Less than 0.01%	2 weeks
Wetlands	Less than 0.01%	1 to 10 years
Biological water	Less than 0.01%	1 week

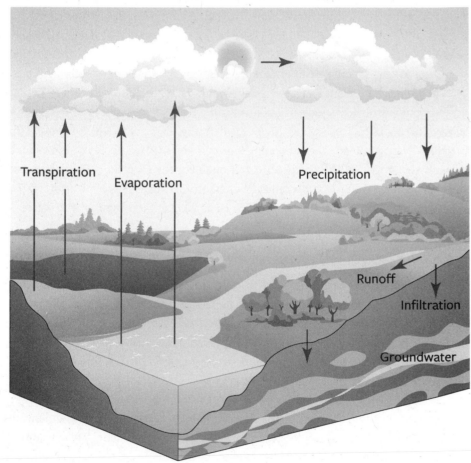

FIGURE 3.1.
The hydrological cycle.

Annually, the globe receives 119,000 km³ of precipitation. This makes the global average rainfall 800 millimeters, of which 480 millimeters evaporates back to the atmosphere and 320 millimeters makes its way to the ocean.

That precipitation comes in the familiar forms of rain and snow but also in the form of condensation, which is often referred to as occult precipitation. Condensing on vegetation, occult precipitation can be the major source of precipitation in some regions. For instance, mist has been estimated to account for 40 to 94 percent of the total precipitation in some highland areas in south and southwestern Africa. Similarly, with coastal regions of the Atacama Desert in South America, condensation accounts for the majority of precipitation.

Deforestation in mountainous semi-arid regions is typically followed by reports of local springs and wells drying up. When these areas are afforested again, consistent reports of springs reappearing and wells filling follow. Though it might seem insignificant, the contribution of condensation to the land should not be overlooked.

The path of water

Knowing how water behaves both on and in soil is important when it comes to deciding which earthworks might or might not be appropriate for a site (see Figure 3.1). The wrong type of earthworks in the wrong soil can fail to do the intended job, wasting money and resources. In the worst case scenario, the wrong earthworks project in the wrong place puts life and property at risk. It is as important to know what earthworks are appropriate for a site as it is how to build them properly.

Let's start with what happens when water first arrives on a site. When water precipitates, gravity is the overseeing force that dictates its movement through the hydrological cycles, though it is not the only one. Water follows the path of least resistance. If you imagine a surface that will allow water to infiltrate uniformly, then water will flow orthogonally to (meaning at 90° to) the contour of the land.

When water infiltrates the soil, its movement is dictated by gravity, capillary action, and the porosity of the soil. The larger the pore space in soil, the more easily water will flow through it. The rate at which water can flow through the soil is known as *hydraulic conductivity*. Table 3.2 shows the hydraulic conductivity of different soils.

When rain first falls on parched soils it might meet an initial resistance to infiltration. This is because of surface tension. Because of the shape of the water molecule, it has a positively charged end and a negatively charged end. The bulk of the molecule's electrons are on the oxygen side, making it slightly negatively charged. The two hydrogen atoms share electrons with the oxygen, making the hydrogen end

Table 3.2. Hydraulic conductivity of different soils.

Soil Type	Hydraulic conductivity at saturation in cm/hour	Soil porosity (fraction of pore space in the total volume of the soil)
Sandy loam	2.59	0.45
Silt loam	0.68	0.50
Clay loam	0.23	0.46
Clay	0.0	0.475

of the molecule positively charged. Along the boundary between water and air, the atoms of water align due to the polar charge the molecules carry. This attraction between water molecules creates the surface tension that can initially retard the infiltration of water. In the case of hydrophobic soils, the soil will actually repel water. Once the soil is wet, however, water can more easily infiltrate, as the surface tension has already been broken. Capillary action can now take place, drawing water into the soil.

This same polar nature of water is also what allows it to remain in the soil and not just drain away due to gravity. Water sticks to most surfaces. If you spray a mist onto a glass pane, the droplets will remain stuck to the glass until they grow large enough for the force of gravity to pull them downward. This is because of a phenomenon known as adsorption. The surface of the glass will have either a positive or negative charge. Because of this charge, molecules of water can form a weak bond (called a hydrogen bond) with the surface. The water is adsorbed to the surface.

This process occurs in soils as well. If we look at a clay particle (Figure 3.2), the hydrogen atoms align to the negative charge on the clay particle, surrounding it with a layer of water molecules. The hydrogen bond is strong enough that it resists many other physical forces. We could say that the molecule is now less interested in things such as gravity, evaporating, or freezing. Swelling clays, such as bentonite, have a nice planar surface. This allows water to adsorb on the surface in a nice orderly fashion. Because of this, there is a field of aligned water molecules, meaning that additional water molecules can be attracted to the adsorbed water molecules, creating additional layers of water covering the molecule. When the clay "swells," the molecule of clay itself maintains its dimensions. It only appears to grow due to the water layers surrounding it.

You will notice that two things happen when soil gets wet. One is that it gets sticky. Wet soil will stick to your hands more easily than dry soil. This is again the

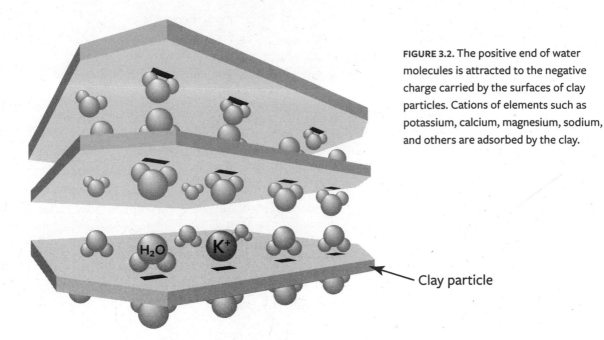

FIGURE 3.2. The positive end of water molecules is attracted to the negative charge carried by the surfaces of clay particles. Cations of elements such as potassium, calcium, magnesium, sodium, and others are adsorbed by the clay.

Clay particle

process of adsorption, allowing hydrogen bonds to hold water to your skin. That water is also attached to soil particles. The other thing you will notice is that it becomes pliable. As water molecules adhere to the surface of soil particles they form a layer, which makes the soil pliable, essentially forming a lubricating layer. This is especially noticeable in clays. If you try to mold dry clay, you will only be able to crumble it. When it is wet, however, you will be able to shape it as you want.

This is helpful for us because there are times when we want to compact soil, such as when building a dam or a clay pond liner. In these cases, having the soil slightly wet aids in compaction. Clay will not compact well if it is dry. If it is damp but not wet—that is to say if the clay particles have a layer of water adsorbed to them but not multiple layers—it will move into position more easily when compacted. If it is too wet, however, the clay will just flow around the compactor. In other words, if it is too wet, it will not compact at all, and machines are likely to get stuck.

Adsorption and surface tension together are responsible for the capillary action in soils. This force that draws water into the pore spaces between soil particles is known as the soil suction. As mentioned above, due to the adsorption on particle surfaces, the water does not behave as it would under normal situations. A smaller capillary will contain less free, or unbounded, water, lowering the freezing temperature of the water in the soil. Experimentation on glass capillaries shows that the freezing point for water in a 1 mm capillary is from –17°C to –21°C, and for diameters of 0.1 mm, the freezing point drops to –25°C to –30°C.

Capillary action is stronger among smaller soil particles, which have smaller pore spaces. A soil such as clay has smaller pore spaces between particles, giving its capillary forces greater power over water movement than a soil with larger pore spaces, such as sand. This gives rise to an unexpected effect. If you have a layer of clay over a layer of sand, water will saturate the clay layer before draining into the sand layer. The capillary force holds the water between the clay particles too tightly to allow the water to follow down into the sand. It is only when the clay layer becomes saturated that gravity will dominate and the water will flow into the sand. Sand has a greater hydraulic conductivity, so the water will drain through this sand layer more quickly than it did through the clay layer. We will see why this is important with respect to swales in Chapter 7.

In the case of a more porous material, such as sand, gravity has more influence over water movement. When sand is located over clay, it will flow down until it reaches the clay. When the water reaches the clay, it can be wicked right into the clay, whether or not the sand is saturated.

Soil is not uniform. Hydraulic conductivity will vary across a section of soil. Not only is pore size important but pore connectivity as well. It is not at all uncommon that subsoils or portions of soil will have poor permeability, which can lead to either puddling or surface runoff if the precipitation rate exceeds the local hydraulic capacity. The same can be said for soil compaction that occurs as a result of tilling or deforestation.

Not only are the soil pores themselves important, but so is their alignment. Pore channels can form as a result of natural alignment of soil particles, biological processes, shrinking and swelling of the soil, and mechanical intervention. The roots of both woody and herbaceous plants mechanically shape soil. Roots push through soil, creating new paths. Old root hairs die off, leaving pockets but also biological material that other life can then feed on. The exudates of plants, fungi, and microorganisms help to bind soil particles, giving healthy soils their friable, crumb-like structure. Arthropods—especially ants and termites—and worms also make significant contributions to soil characteristics. Healthy soils allow water to infiltrate and can hold soil moisture, making water available for biological processes.

Compaction of soils can occur through a number of activities. For example, overgrazing can lead to both mechanical compaction of the soil and loss of plant cover. Bare soils are subject to erosive processes that can cause crusts on soil. The end result you will see in these cases is puddling of water when rains do come. Deforestation also denudes the land, leaving it open to the formation of crusts that increase runoff and puddling and decrease infiltration. In tropical regions this can reduce the hydraulic conductivity of soils by half.

The process of tilling also breaks down the crumb-like structure of soils, decreasing infiltration. In some soils there can be an initial increase of infiltration after plowing. Eventually, however, rain breaks down conglomerated soil particles, leading to crust formation and reduced infiltration.

Hydrology 101

When looking at water movement in soils, we divide the Earth into two zones, one saturated with water, one unsaturated, as in Figure 3.3. The zone saturated with water is known as the phreatic zone (from the Greek root word for *well*). The unsaturated zone, where you will concentrate your work as an earthworks designer is the vadose zone (from the Latin for *shallow*). The barrier between the vadose and phreatic zones is the phreatic line, more commonly referred to as the water table.

The phreatic zone is the zone of aquifers. It is a permeable zone that stores and transmits water. To say that water is transmitted simply means that the water will flow under the force of gravity. In the real world, however, things can and often do get a little fuzzy and relative. Similar to an aquifer, an aquitard (AKA aquiclude) stores water but does not transmit it easily. There is no strong delineation between an aquifer and an aquitard, however. The two are relative to each other. If a section of the ground transmits water more easily than an adjacent section, the two would be considered aquifers and aquitards, respectively. An aquifuge, on the other hand, is solid rock and does not allow the infiltration of any water.

There are two types of aquifers: confined and unconfined. An aquifer can be unconfined, which means that the upper boundary is in permeable soil or rock. This upper boundary is the phreatic line or water table. There are also confined aquifers. In this case, there is an impermeable or less permeable region above the permeable, saturated zone of the aquifer. With the confined aquifer, the upper boundary of the aquifer is geologically determined. With the unconfined aquifer, the upper boundary varies with the amount of groundwater recharge and depletion. The more precipitation that percolates down to the aquifer, the higher the water level. Conversely, the more water that is drawn from the aquifer, or the less water that percolates down, the lower the aquifer will be.

There are also perched aquifers. These are transitory aquifers that arise as a buildup of water above a less permeable zone. You can think of them as a bottle neck, or a phreatic traffic jam. Perched aquifers arise as water passes through the vadose zone to where it meets either an impervious obstacle that it must then flow around, which takes more time and causes water to saturate locally, or it meets a less permeable layer, causing the path of the water to slow enough to saturate the soil. You could think of it as analogous to a sink with a partially clogged drain. When you turn on the tap, water backs up in the sink. When you turn off the tap, the water slowly drains out.

An interesting side note to aquifers are that they are subject to the moon's gravitational pull, as are any large bodies of water. While the magnitude of Earth tides in large aquifers is only 1 or 2 centimeters, it is enough to assist in gas exchange in soils.

Aquifers usually, though not always, play a role in the appearance of streams. In the majority of cases, streams appear where the water table is high enough to reach the surface of a low-lying but sloped area. It becomes easier for water to exit the ground (remember the path of least resistance that water follows) and flow along the surface than to travel under the ground as in Figure 3.4. In these cases, the stream is a discharge area for the aquifer.

There are cases, such as in ephemeral streams in arid and semi-arid regions, where the water flows along the surface in the vadose zone. In these cases, water infiltrates the stream bed, where it can then percolate down to feed the phreatic zone. In these cases, the stream is a recharge area for aquifers.

Lakes, ponds, and wetlands are also typically (but not always) at points where the water table is high enough to escape the ground, making them discharge areas

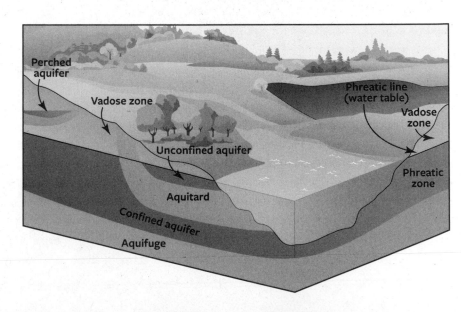

FIGURE 3.3.
The saturated phreatic zone is where aquifers lie. The unsaturated zone is the vadose zone. The phreatic line, or water table, is the upper boundary of the phreatic line.

for the aquifer. It is helpful to understand the movement of water through the landscape.

During a rain event, the majority of the water that falls is either intercepted by the leaves, branches, and stems of plants, where much of it evaporates; is used by plants as biological water; or reaches the soil to later evaporate. The bulk of the water does not reach the phreatic zone.

When water does reach discharge areas such as lakes and ponds, it builds up along the perimeter of the water bodies, pushing the phreatic line upward. This then pushes inward to more closely approximate the profile of the land, and at a higher level than before the rain event. You might then see spring discharges as in Figure 3.4.

Unsaturated ground in the vadose zone transmits water more slowly than in the saturated phreatic zone. As a result, discharge from springs, streams, or standing bodies of water is faster than the flow rate through the unsaturated soil that recharges the aquifer. The relationship between flow rate and water content is logarithmically linked. In other words, as the water gets closer to saturation, the flow rate becomes progressively faster.

This understanding of the movement of water within the ground will be invaluable to you as you design and implement earthworks. Understanding water's movement helps you to better predict the possible outcomes of the various types of earthworks you might choose to put on a site. It can help you to know whether you are going to benefit the site and the local ecology or create a potential hazard that might lead to a failure with a cost in lives and property.

Recharge is faster in areas where hydraulic conductivity is greater. Consider a clay-rich landscape. Clay has a low hydraulic conductivity and does not allow water to infiltrate readily. Clay content decreases with slope, however. The steeper the slope, the less clay content the soil will have. This makes hills important recharge sites. As the water infiltrates

FIGURE 3.4.
The movement of groundwater.

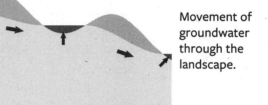

Movement of groundwater through the landscape.

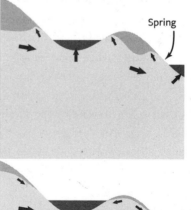

Change in groundwater as rain initially falls.

Spring

Change in groundwater as rainfall continues and more soil saturates.

along the slopes of hills, it has a less resistive route to reach the phreatic zone than in the flatter clay-rich areas. In this scenario, a confined aquifer can form. Planting trees on the hillside—a standard prescription, as you will later see—assists in the infiltration of water. The mechanical action of the roots of the trees, as well as their humus-building detritus and the biological communities that are attracted to the trees, assists in water infiltration.

There are a few types of water flow with respect to groundwater hydrology. The first is overland flow. Overland flow is commonly referred to as runoff. It is water that flows across the surface of the land before reaching a stream or other body of water or eventually infiltrating or evaporating. This occurs when the hydraulic conductivity is less than the rate of rainfall. In other words, it happens when rain is falling faster than it can infiltrate into the soil. (There is, in fact, more to overland flow than this simple model, but for our purposes this model is sufficient.) This is most common in arid and semi-arid regions. A desert flash flood comes about as a result of overland flow. It most often occurs with compacted soil, crusted soil, or hydrophobic soil. Overland flow is not a common occurrence in humid temperate regions.

What is more common in humid temperate regions is saturated overland flow. Saturated overland flow occurs when water infiltrates the vadose zone until it temporarily reaches the saturation point and then escapes through the surface, similarly to the appearance of streams in areas of high water tables. Remember that saturated regions have the greatest flow rate and are the path of least resistance, at least until the surface becomes an easier path for water flow. The bottom, concave sections of hillsides are most prone to saturated overland flow. If these wet areas are present at the base of hills, water-harvesting earthworks that capture and store more water in the ground will only make these areas wetter.

Throughflow, or lateral flow, is the water that flows through the vadose zone near the surface. If it temporarily reaches the saturation point, it is responsible for the saturated overland flow mentioned above.

Finally, there is groundwater flow, which is the flow in the saturated phreatic zone.

Bare soils have lower hydraulic conductivity than covered soils. Both vegetated and mulch soils prevent the formation of soil crusts that can retard infiltration. Soil crusts will make water more likely to be lost to evaporation. While it is true that water intercepted by plants and mulch then evaporates, the soil structure remains intact, and the water is temporarily available for biological processes. With bare earth, this is not the case.

The layout of landscapes

Building earthworks always involves working on water's interaction with gravity. This means that elevation changes are a factor. There are approaches to water harvesting that can be done on sites that are essentially flat, but most locations involve working with hillslopes.

Soil is composed of aggregates of organic and inorganic particles. For the inorganic mineral particles, the smallest particles, of less than 2 µm, are clay particles (known as soil colloids). One size up from clay is silt, at 2 to 50 µm, followed by sand, at 50 µm to 2 mm. Particles are gravel if they are from 2 mm to 64 mm in size. Cobble is 64 mm to 256 mm, and anything larger than that is a boulder.

Every part of a hillslope experiences shear forces, which are forces in which two media are laterally in tension against one another. In the case of an individual soil particle on a hillslope, gravity pulls it downward. It cannot move straight down because it is resting on the hillslope with other soil particles underneath it. Therefore it wants to slide down the hillslope. Frictional forces from the other particles hold back the shear force on the particle, preventing it from sliding downhill. If the shear forces are greater than the frictional forces, down it will go.

For every type of soil, there is an *angle of maximum stability* at which point a slide is triggered. The name is a little confusing because it is actually the point at which a pile becomes unstable. It occurs if you pour particles in a pile. The particles will build up more and more steeply until the shear forces are too great and a slide occurs at the angle of maximum stability. The angle the particles slide to is called the *angle of repose*. This is considered the steepest stable slope for that type of particle.

Water can play a role in the angles of maximum stability and repose. Recall how water adsorbs to soil particles. The polar bonds in water also cause surface tension. If you wet soil slightly, but not to saturation, the soil particles still touch one another, causing friction. This friction overcomes shear forces. The surface tension of the water further binds the soil together. This is how you can make a vertical wall when building a sandcastle. If you saturate the sand, however, the individual grains of sand can lose contact with each other, lowering frictional forces below the shear forces. The sand can then flow like water in a process known as *liquefaction*. If you build a vertical wall on a sandcastle and then steadily add water to that wall, it will eventually undergo liquefaction and collapse. The same thing can happen on hillslopes, which is why care is needed in determining whether earthworks should be added to a site and, if they are added, what type is safe to use. Hydrating the wrong hillslope can trigger a slide. We will look more closely at landslides in Chapter 9.

One recent scientific discovery of note regarding uniformly mixed soil particles of differing sizes is that slides can cause the different particles to separate into alternating stripes of each particle type, with each stripe being of roughly uniform width (Makse and Stanley, 1997). Should you be excavating a hillslope and find alternating bands of two different particle types (sand and silt, for instance), it could be evidence of past slides.

Hillslopes can be divided into sections. Traditionally, permaculture looks only at what is termed the "keypoint," after the work of P.A. Yeomans. (More on this model in a moment.) The keypoint is the point on the land where the profile of the hill changes from convex to concave. It is helpful to look a little closer at hillslopes, however.

Hillslopes have a convex top, a straight section in the middle, and a concave section at the bottom. On some landscapes, the straight section is essentially absent, and the land changes from convex to concave with no straight run. The upper concave section is the region of erosion and may or may not contain a free face, which it to say it may contain a scarp or cliff. Material erodes from here and falls to the straight section, where deposition of material starts. The concave section is where the pediment deposits. If you track the erosion of a hill over millions of years, it will start with a steep profile of rock and gradually erose down to a stable hillside.

Using the hillslope units devised by Robert Ruhe (see Figure 3.5), the top of the hillslope is termed the *summit*, which falls into the shoulder. The convex *shoulder* may drop into a *free face* or may transition into the straight *backslope*, which then falls into the concave *footslope*. The trailing end of the concave section is the *toeslope*. We will use these terms in reference to the parts of a hillslope throughout.

It's typically recommended in permaculture texts that the base of the backslope (keypoint) be the highest point at which earthworks are carried out. The reason for this is that the areas above the backslope are more erosive. Not only are earthworks here subject to more erosion, they can also trigger more erosion. Building earthworks involves disturbing the land. Removing ground cover and digging in the convex region of the shoulder can increase erosion. In the case of dam building, several obstacles make the shoulder a bad location. Firstly, the catchment is not as great and might not be enough to fill the dam. Another issue is the material needed to build the dam wall. Because of the profile of the land, the amount of the type of material needed to construct the dam wall is greater in the convex shoulder than on the backslope or footslope.

In the landscape model for open drainage systems devised by P.A. Yeomans, a landscape has a main ridge, from which *primary ridges* fall. In these primary ridges

Summit

Shoulder

Free face

Backslope

Keypoint

Footslope

Toeslope

FIGURE 3.5.
Slope units developed by Robert Ruhe. The keypoint, used in permaculture, is located at the transition between the backslope and the footslope.

Yeomans described *keylines* in the head of *primary valleys* made by the adjacent primary ridges. The keylines extend from the keypoint (at the base of the backslope) in the primary valley. In geomorphology, primary ridges are known as *nose slopes*; the keypoint rests in the *head slope*; the slopes between the head slope and the nose slope are *side slopes*; the bottom of the system is known as the *base slope*; and the top of the system from which material is eroding is known as the *interfluve*. (See Figure 3.6.) The head slope and side slopes create a *convergent* area that is wetter, whereas the water is directed away from the nose slope, making it a *divergent* area that is drier.

As you move higher and higher up a main ridge, the backslopes (and keypoints) of each head slope are generally higher than their adjacent head slope down the main ridge. Erosion concentrates in the valleys with water flow. As a result, the nose slope is generally not as steep as the head slope.

Hillslope-channel coupling, a concept from geomorphology, is useful to the design of water-harvesting earthworks. In geomorphology, a hillslope is coupled to a

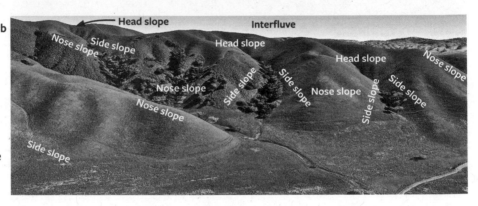

FIGURE 3.6.
(a) Diagram shows the classifications of landscape used by P.A. Yeomans in keyline design.
(b) Diagram shows the classifications used in geomorphology.

stream channel if erosive sediment spills into the stream channel. We can think of it as how connected a site is to its nearest stream.

We can better avoid overharvesting water if we think in terms of the water we prevent from reaching stream channels. The more coupled the landscape, the greater the chance of overharvesting. It should be noted that overharvesting is an issue in some areas. Just as we saw with the case of the Colorado River in Chapter 1, inappropriate and excessive upstream claiming of water has the potential to have devastating downhill effects.

The final landscape concept we'll examine is the Strahler stream order. The Strahler number was devised by Robert Horton and Arthur Strahler to order tributaries in a waterway based on their size. At the higher regions of a watershed, headwaters appear. These are first order streams. When two first order streams meet, they form a second order stream. Then two second order streams meet, they form a third order stream, and so on. In this model, a first order tributary meeting a higher order tributary does not form a new, still higher order stream. In other words, if a first order stream meets a third order stream, the third order stream remains a third order stream. You are most likely to work on a watershed with a third order stream at the terminus. The world's largest is the Amazon river, which is a 12th order river at its mouth.

References

Davie, Tim. *Fundamentals of Hydrology*. Second edition. Abingdon: Routledge, 2008.

Ghosh, S.N., and V.R. Desai. *Environmental Hydrology and Hydraulics: Eco-technological Practices for Sustainable Development*. Enfield: Science Publishers, 2006.

Goudie, A.S. *Encyclopedia of Geomorphology*. Volume 1 A-I. London: Routledge, 2004.

Huggett, Richard John. *Fundamentals of Geomorphology*. Second Edition. Oxon: Routledge, 2007.

Lal, Rattan (ed.). *Encyclopedia of Soil Science*, Second Edition. Boca Raton: CRC Press 2006.

Makse, Hernán, H. Eugene Stanley, et al. "Spontaneous Stratification in Granular Mixtures." *Nature* 386, (1997) doi:10.1038/386379a0.

Martin, R. Torrence. "Adsorbed Water on Clay: A Review." *Clays and Clay Minerals*, vol. 9, issue 1 (1960) doi: 10.1346/CCMN.1960.0090104.

Park, Chris. *A Dictionary of Environment and Conservation*. Oxford: Oxford University Press, 2008.

Shukla, Manoj K. *Soil Hydrology, Land Use and Agriculture: Measurement and Modelling*. Wallingford: CAB International, 2011.

Todd, David Keith, and Larry W. Mays. *Groundwater Hydrology*. Third Edition. Hoboken: John Wiley & Sons, Inc., 2005.

United Nations Environment Programme (UNEP). "A Glass Half Empty: Regions at Risk Due to Groundwater Depletion." 2012-01. na.unep.net/gcas/gctUNEPPageWith ArticleDScript.php?article_id=76.

Designing for the
Whole Environment 4

Permaculture and sustainability

Water-harvesting earthworks are frequently employed in permaculture. If you are not familiar with a formal definition of permaculture, it is a method of designing sustainable systems to meet human needs. The focus on sustainability is what sets permaculture apart from other design methods. Unfortunately, the word *sustainable* is usually used without a clear definition. To really comprehend what sustainability means, we need to bring clarity to the term.

Most dictionary definitions hint at human action carried out in a way that does not completely deplete or destroy natural resources. This definition is still imprecise enough that the term gets thrown around in instances when it is not applicable. Permaculture involves designing ecosystems to meet our needs. Any system built, whether mechanical or biological, will have an initial cost of construction and generally ongoing maintenance costs to keep the system operating. If the costs outweigh the returns of the designed system, that system is not sustainable.

The metric we use to measure the costs and returns of a system is energy. If a system captures and stores more energy over its lifetime than it uses, it is sustainable. Consider a swale. Swales are water-harvesting ditches built on contour so that they can catch runoff and allow it to sink into the ground. If you spend two hours with heavy machinery, burning diesel fuel, to make swales in gravel-dominated glacial till where there is and never can be any runoff, the whole project is a waste and cannot be sustainable. If, however, you spend the same two hours to construct swales on a site with significant runoff that then allows you to capture and store enough rainwater to establish trees, the system has a chance to be sustainable. The trees can produce not only timber, fuel, and possibly food; they also produce nutrients that feed the soil and create habitat for other organisms. This biological activity from the newly established ecosystem further contributes to the fertility of the site. In other words, the right earthworks in the right place can significantly contribute to the health of the system. The initial energy costs are repaid with the energy the system captures.

Permaculture has three ethical guidelines to help us ensure sustainability in our projects. These guidelines keep us on track, and help us avoid making choices that harm our future and the health of the planet. They are:

1. care of the Earth
2. care of the Earth's people
3. return of the surplus of our productivity to the Earth and its people

The second guideline reflects the fact that we are designing systems to meet human needs. We cannot do this sustainably, however, if we damage the ecological system that makes life possible. Hence the first guideline, which directs us to take care of our life-support system. The third guideline reflects the fact that every living system is a part of its environment. Materials flow through the system to make life possible. If we want a healthy planet, we need to allow the Earth's nutrient cycles to flow to maintain life. If we want healthy societies, we need a similar flow of resources through those societies. Living systems are flow systems, and in flow systems, stagnation is death.

The design process

Any good design will consist of three interconnected stages: goal setting for the project, the actual project plan, and adjustments made in the face of feedback from the system. The goal-setting stage is the single most important stage and is hardest to do well. Everyone who starts developing an earthworks project has some desire in mind, some reason they want to design and build earthworks. The problem is that very often the desire does not significantly express the motivation for the project. Every desire beyond your immediate physical needs has a deeper motivation. To create a good goal, one that will lead to an outcome you will be satisfied with, you will need to identify the deeper motivation behind your desire. On more than one occasion, I have had clients who were either permaculturists or permaculture enthusiasts who consulted with me about the design and installation of swales on a site when their underlying assumption was that permaculture sites need swales. Their motivation was to be good permaculturists, and for them that meant there were swales on site. As it happens, in each such case it turned out that swales were either unnecessary or dangerous on their site.

The task of uncovering your underlying goal is done by asking why you have the desire for a certain course of action. What need are you trying to meet? Is there a personal value you are trying to express? Will this project actually meet your needs or live up to your values?

A good goal is worded openly enough that it does not prescribe a set course of action. If your goal prescribes a specific course of action, it is a plan, not a goal. In that case, you need to stop and spend more time figuring out in more basic terms what it is you are trying to do. Consider, for instance, having a goal to "make better use of available rainfall to make our cattle operation more resilient." Contrast that with "build swales and dams on site to capture the available runoff for our cattle operation."

The first goal provides flexibility with respect to the planning of the project. Essentially every option is open with this goal, and you are free to choose the most appropriate path for your site. If something does not fit your particular site, you can simply choose a better option.

The second goal is committed to a specific action plan from the beginning. Whether swales and dams are perfect or potentially disastrous for the site and the environment surrounding the site, they are the prescription. There is no alternative course of action—it's swales and dam, or the project is a failure. That goal is already a plan.

When your goal is sufficiently expressed, you are ready to move on to the planning phase. There are two things to know about plans. The first is that plans are wrong. You cannot and will not make a perfect plan. You will take imperfect information and use your fallible human mind to craft a plan. Perfect plans are impossible. We need to recognize uncertainty and our own human limitations when it comes to planning. This leads to the second point about planning. Never fall in love with your plan. The quick and easy path to disappointment, and too often disaster, is to fall in love with your plan. I recommend holding your plan in suspicion. Know that there are flaws in there somewhere, even if you don't see them.

Planning decisions should be checked against your goal. The goal acts as a decision-making engine for your plan. When you consider a course of action, see whether it falls in line with your goal, violates your goal, or is superfluous to your goal. Options tend to arise during the course of planning or implementation of the plan. Checking with your goal can help prevent you from being sidetracked or incurring unnecessary expense. For example, it is not uncommon to become excited at the prospect of expanding a project once the machinery is on site. It can be tempting to think that, while the machinery is there, it makes sense to make everything bigger or to install more water-harvesting elements. Additional machine time means additional expense. Referring to your goal in the face of such temptations can help prevent you from overspending.

Throughout the goal-setting stage and planning stage, you adjust to feedback. Information you receive can change your plans or even your goal. Your plans, and even your goal, are imperfect. The feedback you receive is from the real world. Defer to the reality of the situation rather than trying to bend nature to fit your expectations.

Permaculture strategies

Permaculture uses a number of strategies to develop sustainable systems. These help to make the designed systems operate more efficiently and with more resilience. The first such strategy is that of *stacking functions*. This means that each element placed on a site is intended to have at least two functions. For instance, a dam can serve as a backup irrigation system for a farm. That dam can also be used for aquaculture, now giving it two functions. Dams are also very effective at attracting wildlife, and wildlife carry in with them significant nutrients, making the dam a site for nutrient accumulation. If we intercept water flowing through the catchment, we have a responsibility to use that water wisely. By making the water that we capture carry out multiple tasks, we are maximizing its potential, and being better stewards of the land.

Another useful strategy is *functional connectivity*. This refers to making beneficial connections between two or more elements. For example, a swale can be a useful way to capture runoff water. The swale can be integrated with a dam to direct water into a dam for open water storage. The swale can also have a level-sill spillway built into it to safely discharge excess water in the dam. As the dam fills, it will backfill across the swale, where the water can infiltrate the ground and increase groundwater storage. The spillway can be located such that the overflowing water discharges in a position where it will have a maximized path length before it leaves the property, giving it greater opportunity to infiltrate the soil.

The final permaculture strategy we will examine is *redundancy*. Redundancy adds to site resiliency. All water-harvesting earthworks are essentially backup systems for irrigation. A site without water-harvesting earthworks either relies on available precipitation and the site's existing hydraulic conductivity to provide water or uses an irrigation system that is dependent upon pumped up groundwater or water imported from off site. Another example would be the recommended dimensions for earthen dams. While it is possible to make dam walls smaller, the methods recommended have redundancy in mind to help ensure a strong and secure dam wall.

Source to sink

If we look at the local hydrological cycle for an individual site, water travels from a source—either falling in the form of precipitation or entering the site from a stream or a spring—to the sink, which is where it leaves the site.

To take best advantage of the water on site, we want to slow its path from source to sink. This gives us more opportunities to put it to productive use. To slow its path, we can lengthen the path, which might involve either making the above-ground path longer or sinking water into the ground, or we can slow the path by intercepting and storing the water. It is important to note that a site benefits the environment if it slows the water but does not halt it. Blocking all water from moving off site robs every point downstream of the productive potential of that water. We need to remain mindful of the downstream effects.

When it comes to planning and installing water-harvesting earthworks, start as high in the landscape as is practicably possible. You can leverage a lot of potential out of the highest point of land for four reasons. Firstly, the scale of the earthworks will be much smaller than at lower points in the landscape. At the top of a local watershed there is less runoff than at lower elevations. Secondly, by encouraging water infiltration at the top, you are maximizing the groundwater recharge potential. Recall the global trends for groundwater from Chapter 1 and the typical hydrology of streams from Chapter 3. Getting water into the ground early means that the groundwater has a longer, slower path than it would were it to infiltrate lower in the landscape. Hillslopes contain more surface area, and usually volume, of soil per acre (which is measured as a level two-dimensional area and does not account for vertical land area) than the same area on flat ground. This gives the hillslope more potential to hold water in the vadose zone. In this respect, you can think of the hill as a large, leaky bucket. Under the right soil conditions, it makes sense to direct water into that bucket. When this is done, it is common to see higher water levels in downhill wells, greater flow durations in downhill ephemeral streams, and more consistent flow volumes in downhill streams. Thirdly, recall that the top of hillslopes is where erosion takes place. Should you capture erosive runoff high in the landscape, you can reduce soil loss to erosion. And fourthly, the soil catena of hillslopes is such that clay becomes less prevalent with increases in incline, as smaller particles are more easily washed downhill. Steeper slopes typically mean lower clay content, which in turn means greater hydraulic conductivity in the soils in hillsides. This makes hillslopes ideal locations for groundwater recharge as it is typically easier to get water to infiltrate in the summit, shoulder, and backslope than in the footslope or toeslope.

Reading the land

Designing appropriate water-harvesting earthworks is a difficult challenge. Fortunately, there is a cheat sheet that makes the process easier, and that is the land itself. The best dictator of what should be done on the land is the land. The first place to start is pre-development history. What was the land like before the forests were cleared for settlement? Were there, for example, many beaver dams across the landscape, making many small open water storages? Try to find research on the historical ecology of the region. What were the traditional human impacts on the land? Indigenous land use might give you a clue as to what will help or hurt a site. Amazonian land practices tended to build healthy, resilient systems, whereas ancient Mesopotamian forest clearing caused massive erosion and ultimately desertification.

The health of the land at its peak environmental health is the minimum standard you should aim for. If the land was a dry savanna that later desertified, then a savanna should be the minimum standard you have for the site. You should aim to restore the land to its peak historical health and surpass that historical peak in terms of resilience and biodiversity, if possible.

For cleared sites that were historically forested land, the cycles of flood and drought will have changed. The presence of forests has a complex hydrological effect on the land. Floods and droughts exist in forested lands, but the clearing of lands makes the cycles of flood and drought more severe. As you will see in Chapter 7, trees increase hydraulic conductivity. The increased infiltration as a result of forests means greater residency time for water on the land. With more water in the ground, streamflow becomes more consistent. Given the same volumes of rain, total flow of streams will be much greater in a catchment denuded of trees, but this increase flow happens in periods of high volume that correspond with rainfall. In other words, if you clear the trees, the water runs off the land quickly. Along with this will be decrease groundwater recharge, loss of soil fertility, increased erosion, more severe drought periods, and a decrease in biodiversity, biotic activity, and overall resiliency of the land. Over time, the loss of forests will also lead to a decrease in rainfall. There is more to the health of riparian systems than the single measurement of streamflows.

Next, carefully observe the site. Are there areas where water pools in heavy rains? Areas that dry out easily? Evidence of surface flow or erosion? These are all clues as to where you can effectively concentrate your efforts. Study the topography of the site. How is water forced to move due to the lay of the land?

Before you start planning any earthworks for the site, look for potential dangers and restrictions. Does the land show any evidence of landslides? Adding water-harvesting earthworks on unstable slopes can lead to catastrophe. Are there unstable soil types on site? Saturating the wrong type of soil can trigger landslides on land that appears flat. What is downhill from your site? Make the assumption that whatever you build will suffer a catastrophic failure when the water level is at maximum capacity. Is there a potential for loss of life or property? If so, then you will need to either change your plan or establish some sort of failsafe to protect downhill areas. See Chapter 9 for a more comprehensive look at safety.

The potential for overharvesting

Water is an issue that starts conflict, and greater warfare over water is a major concern in a world where the demand for water is increasing at twice the rate of population increase. The issue becomes even more sensitive in semi-arid and arid regions.

Permaculture was born in Australia, and many of the water-harvesting earthworks employed by permaculture were and are widely used there. Since the 1970s, there has been an increase in the deployment of farm dams. Since that time, there has also been a recorded decrease in rainfall of 15 percent. These two factors have led to a decrease in streamflows throughout the nation. To make matters worse, the impact of dams on streamflow is greater during dry years than during wet ones. This has, understandably, led to conflict. As a result, the establishment of new dams is coming under stricter regulation.

In general, regulations around water-harvesting earthworks exist for two reasons. Firstly, water is a public good. You have a right to water on your site, but so do people downstream. Water is a public good, and if you claim it as your private property, you are running afoul of both the law and universal human rights. As seen in Chapter 1, claims to water upstream have destroyed entire ecosystems. Water is yours to make use of, then pass on for others.

A study of the Yass River in the Murray–Darling Basin (Neal et al., 2002) found that for every 1 megaliter (264,000 US gallons) of farm dam storage there was a 1.3 megaliter (343,000 US gallon) reduction in streamflow. Similar reductions have been found in other studies, as well. This reduction in streamflow increases the duration and occurrence of periods of low streamflow and zero streamflow. That the rate of reduction of flow is greater than the volume of the dam is not surprising. Not only is the captured water used for irrigation of crops or livestock, there are also significant losses in dam volume due to evaporation. In humid areas of Victoria, 10 to 20 percent of a dam's volume is lost to evaporation, and that figure jumps as high as 70 percent in the drier areas of the state (Melbourne Water, 2016).

Dams are associated with a lowering of streamflows, though the problem is more complex than simply the introduction of dams. For the sake of agriculture, much of the native eucalyptus forests have been cleared. In one study looking at catchments in the Kent River in Western Australia (Callow and Smetten, 2009), just 36 percent of the native vegetation remained in the study area. The rest had been cleared, largely since the 1940s, with the land now used for the production of annual cropping. This is significant in several ways. The clearing of forests reduces rainfall and occult precipitation. (See Chapter 7.) It sees a reduction in hydraulic conductivity, which means greater runoff of the rain that does fall. Farm dams established for crop irrigation are drained down in the service of crop irrigation. This means that dams are often emptied, so more of the total annual precipitation is taken out of the local hydrological cycle. By contrast, dams for livestock irrigation are not drawn on as much and have a tendency to remain fuller than dams for irrigating crops. This means that less water is used and more water overflows in the spillway to return to the watershed. The impact on streamflows for livestock dams is, in general, less than that for crop-irrigation dams.

The situation around dams is not cut and dried, even for more sensitive drylands. As you will see in Chapter 6, check dams and even small-scale concrete dams in semi-arid regions can help to re-green areas. Streamflow volumes are one measure of what is happening in the hydrology of a landscape, but it is not the total picture. Semi-arid Andra Pradesh, India, was traditionally a tropical monsoon region, but biotic pressures have reduced the land to a semi-arid state. The major rain events come during the monsoon season. With the land as barren as it is now, there is tremendous runoff that very rapidly rushes off the land, through river ways, and out to sea. The monsoon rains make up the bulk of the water available for both people and wildlife to use through the course of the year.

Adjacent to a site highlighted in the Indian case study in Chapter 7 was a small concrete barrier dam holding around 35 million liters (9,246,000 US gallons), built by the Rural Development Trust, a local NGO. This dam unquestionably reduces total streamflow volumes. It also increases residency times for water in the landscape. The base of the dam has a steady throughflow, which has made possible the establishment of a mango orchard and rice paddies that employ dozens of local townspeople. While it is true that barrier dams have a greater impact on river systems, the scale of this dam is quite small, and it is high in the watershed. Additionally, its long and narrow shape is sheltered between two hillsides, reducing wind exposure, which in turn reduces evaporation. Yes, it is reducing total streamflows, but it is providing a livelihood for the neighboring hamlet, and it is an important refuge for local wildlife.

In both Andra Pradesh and Australia, the greatest impact on the hydrological cycle has been the removal of forests. The source of the water problem is deforestation. Blaming the increasing aridity solely, or even mainly, on water-harvesting systems is ignoring the major culprit. Indeed, earthworks can be a beneficial, or even vital, ingredient in reforestation efforts. Inappropriate small-scale earthworks can compound water problems but are not the sources of them. Increased development of silvopastures (systems combining trees and pastures) for livestock operations and alley cropping for annual crop production need to be increasingly employed in drylands in order to combat aridity.

Dams can decouple catchments from the watershed, which contributes to reduction in streamflows. Dams make a contribution to groundwater, though this is not well studied, and it is reasonable to expect that the contribution is not very significant. For a storage dam, effort is put into minimizing leakage, meaning that infiltration of water into the soil is discouraged. In the case of traditional India *johad*, however, infiltration is the goal. In a *johad*, runoff is intercepted and held in place specifically so that it has time to infiltrate the soil and recharge groundwater. There is a similar outcome in the natural world, in which beaver dams contribute to groundwater recharge. Though the beavers' motivation is not to recharge groundwater, recharging is nevertheless a byproduct of their damming of small streams. Higher groundwater levels are recorded as a result of beaver dams, which contributes to the consistency of streamflow downstream. Depending on the design of a dam, there is potential to make significant contributions to groundwater, which can positively affect streamflow.

It is worth mentioning, however, that although the usefulness of dams is greatest in times of drought, these are also the times in which they have the greatest impact on the hydrological cycle. If, for example, your dam is three quarters full, and your dam's catchment collects half the dam's volume in rainfall, then the excess one quarter flows downstream to others over the spillway. If, however, the dam is only one quarter full, and the same volume falls, all the water will be captured. The impact of dams is greatest in drier times.

An effective water-harvesting system that maximizes water availability, both on site and downstream, will seek to slow, not stop, the flow of water. Infiltration is a prime strategy in this approach. This makes water biologically available for soil life and for plants as well as recharging groundwater, making more water available in wells and providing better flows in springs. Streamflow is important, but it is a deceptive measure of riparian health. Maximized streamflows are not, and should not be, the goal. If you completely denude a landscape of all vegetation, runoff will maximize. As a result, streamflows will increase as well. In this scenario, stream-flow volumes will fluctuate wildly with precipitation. Denuding enough of the land-scape will lead to decreases in rainfall and wilder streamflow fluctuations. Runoff water will also be warmer, reducing its dissolved oxygen content, and will carry more sediment. This approach has been inadvertently used time and time again in human history. The floods of ancient Sumer and the gradual desertification of Mesopotamia are our first recorded instance of this, though the pattern has been repeated throughout the world. Clearly, maximizing streamflow is not the goal. Rather, we should seek for consistency of streamflow.

Water-harvesting earthworks are neither a panacea nor a scourge. The effects of hydrology on ecology are nonlinear and are subject to complexity. This means that we cannot fully predict what the outcome will be based on the initial inputs of the system. There is no simple, linear equation that says dryland areas plus dams equals downstream drought, or success for that matter. The human mind has a tendency to gravitate toward dichotomy. We notice a success with earthworks, and they become good things in our mind. Or we see a negative outcome, and earthworks become harmful human intrusions on the land. Landscape hydrology and ecology are too complex to make this kind of reduction. This is why the design process is so important. We need to develop responsible goals for our projects, then adjust to the feedback available. The first guideline for permaculture is to care for the Earth. Our goals need to be aligned with this guideline, and we need to check feedback for the impact of our projects to ensure that they are living up to that guideline.

Permitting and legal restrictions

What earthworks you can install are limited not only by the natural conditions of your site and its surrounding environment and watershed but also by legal restrictions. Various governmental bodies, from national to municipal, may and probably do have regulations governing what you can do on your site. Often regulations are more lax on farms, allowing for greater freedom in water harvesting and storage. The intent of regulation is to ensure safety for the public as well as the individual.

Regulation is also there to ensure that water, a public good, is distributed fairly. In other words, they want to avoid the problems associated with landscape decoupling and reductions in streamflow.

Before you develop a site plan, make sure you know what is legally permitted. You will save a lot of time and frustration if you know up front what you are permitted to do and how it is to be done. For example, in some places, you are allowed to harvest a maximum of one fifth of the water that falls on a given site. Some places require an engineer for a dam of any sort, including farm dams. It is better to know these restrictions before you make a plan for the site.

Working with crews

Even if you are a commercially experienced heavy equipment operator, your crew can give you valuable insights when it comes to the installation of earthworks. Most operators have dealt with nightmarish customers and contractors, so when you first start a job they will tend to be a little skeptical of you until they know you. For your part, you can explain to them what it is you are trying to do and how you think it should be done. Let the crew know that you recognize their skills and that you want to hear their input on the best way to carry out the job. There are lots of site logistics to work out, including where to start, where to place excavated material, and what areas of the site might be problem areas. They know how to move earth around. If you don't trust them to do this effectively, you have the wrong crew.

There are some areas in which you might need to assert your expertise. Particularly if you are doing something the earthmover has never done before, you might need to insist they do something a way quite differently from what they are used to. For instance, dam building in temperate areas is something that most operators have no experience in. For the sake of speed and cost minimization, they may want to try to compact too much earth at one time (for example compacting 60 cm of material at a time instead of 15 cm). This would be the typical method when constructing clay barriers for a garbage dump, but it will be inferior for dam construction.

Whether you are installing earthworks on your own site or on a client's site, you are the person responsible for overseeing proper construction. Keep your eye on the work. Regularly check the material that is being excavated and used, particularly when dam building. And if something is not right, interrupt the operator and correct it. Major projects such as dams cost tens of thousands of dollars. While you need to be respectful of your crew, don't be timid or concerned about appearing critical when the success of the project is on the line.

References

Callow, J.N., and K.R.J. Smetten. "The effect of farm dams and constructed banks on hydrologic connectivity and runoff estimation in agricultural landscapes." *Environmental Modelling & Software*. Vol. 24, Issue 8. (2009) doi: 10.1016/j.envsoft.2009.02.003.

Hill, A.R., and T.P. Duval. "Beaver dams along an agricultural stream in southern Ontario, Canada: Their impact on riparian zone hydrology and nitrogen chemistry." *Hydrological Processes*, 23, 2009.

Majerova, M., B.T. Neilson, N.M. Schmadel, J.M. Wheaton, and C.J. Snow. "Impacts of beaver dams on hydrologic and temperature regimes in a mountain stream." *Hydrology and Earth System Sciences*, 2015.

Melbourne Water. "Diamond Creek: Local Management Plan." 2016 melbournewater .com.au/waterdata/waterwaydiversionstatus/Documents/Diamond%20Creek%20 local%20management%20rules%202016.pdf.

Neal, B., R.J. Nathan, S. Schreider, and A.J. Jakeman. "Identifying the Separate Impact of Farm Dams and Land Use Changes on Catchment Yield." *Australian Journal of Water Resources*, Vol. 5, Iss. 2, 2002.

Elements of
Design and Implementation 5

This chapter addresses site selection, mapping, and observations, as well as the machines used in earthworks implementation. Deciding which earthwork type goes where depends on the soil type and hydraulic conductivity of the soil, runoff volumes, catchment areas, the slope of the land, the climate of the site, and your intended outcome.

Rain volumes

Start with the basics for your site. Check the weather records for the region. What is the record rainfall, and at what time of year did it occur? What is the record drought? How frequently do large rain events occur? Do these follow a seasonal pattern, such as the hurricane season? What is the mean recurrence interval for large rain events and for droughts? What is the variation in total precipitation between a dry year and a wet year? If it is a cold climate, what is the typical contribution from the spring melt?

The size of the earthworks you build will depend in part on the volume of rain you can expect in a large event. You may also face legal restrictions regarding the amount you are allowed to capture. Safety is also a factor in the design of open water storage. The larger the storage, the greater the potential damage it can cause if it fails.

Climate effects

Climate is a major determining factor in both the scale and type of earthworks you will employ on a site. Irrespective of your climate, water security will be a primary goal. In an arid region, you are more likely to take the Nabatean approach of capturing everything you possibly can. (See Chapter 2.) In a wet tropical environment, you generally want to have a backup for dry periods, establish aquaculture systems, or even control excess water. Temperate systems tend to sit happily in the middle, with the systems typically intended to provide backup irrigation or aquaculture. Arid and semi-arid climates need to focus on capturing as much water as possible while minimizing evaporative losses. Making the correct match for your climate will result in more favorable outcomes for your water-harvesting systems.

Temperate climates

Apart from alpine regions in equatorial zones, temperate climates are characterized by clearly delineated seasons governed by the Earth's axial tilt. Westerly winds are usually the prevailing winds, whether in the northern or southern hemisphere, and the weather is usually controlled by ocean-driven cyclonic systems. This means that while the rainfall contributions from trees, noted in Chapter 7, do affect precipitation, the effect is less pronounced in temperate regions.

In temperate regions, the winter is usually the season with the greatest precipitation, with dry periods occurring in the summer. Droughts are a normal occurrence and regularly affect agriculture. Rainfall is often ample in the spring and can be strong enough to delay spring planting. It is in the middle to late summer that drought tends to occur; however, spring droughts have become more common in some regions as climate change worsens.

Strategically, you will want to capture and store precipitation during the wetter periods of the year and hold them in reserve for the drier periods. In colder regions, the spring and fall will provide greater rainfall, with winter precipitation being mostly in the form of snow. Strategies to store spring runoff will help with water security during the summer. This may come by way of either open water storage or by infiltrating water for storage in the soil. As weather is becoming more erratic, and severe droughts are occurring with greater frequency, the sound strategy is to design around the assumption that you are preparing for the worst recorded drought.

In more humid temperate climates, evaporative losses will be less of a problem than in drier regions. It may still be in your best interest to try to minimize evaporative losses by planting trees around the perimeter of reservoirs (remembering never to plant trees on dam walls) or to employ the floating gardens described in Chapter 8. This approach has the potential to greatly expand site production, and the microclimates established by open water storages can extend the growing season.

Bill Mollison has stated that, in humid regions, up to 20 percent of the surface area of a site can be devoted to open water storage. The actual amount of land dedicated to open reservoirs will be dictated by the geography of a site, regulations governing water usage, and budget. In temperate environments, cisterns are generally limited to collecting water off of roofs, though the opportunity to harvest runoff in a cistern may present itself. Dams and ponds tend to be the preferred open water storage techniques.

For interception and infiltration of water, swales and ripping are the typical approaches used. Recall that runoff tends to be less in humid climates. As a result, swales will not be as large as they would be in drylands. In many cases, runoff volumes are insignificant, and swales will actually fill with throughflow rather than runoff. In such a case, the only practical function for a swale would be to fill a dam. Given the correct soil type (see Chapter 7), ripping will often be a more effective water-harvesting approach. Finally, trees should be planted as part of a water-harvesting strategy wherever possible. Alley cropping with coppice and pollard trees can be used for crop production. Livestock operations will benefit from silvopastures. (See Chapter 7.)

Aquaculture can be carried out as either a primary or secondary function of open water storage. Aquaculture in temperate climates is generally limited to raising fish and crustaceans. Options for growing aquatic plants for human consumption are limited in comparison to tropical climates. While there a number of wild aquatic edible plants, they don't carry with them the breeding that domesticated plants do to make them both more productive and more palatable. Efforts to domesticate these plants hold the potential to expand temperate aquaculture. The two most common commercial aquatic plants in temperate climates are cranberries and hardy rice varieties such as Nanatsuboshi, Yumepirika, and Oborozuki, which are grown in the cold climate of Hokkaido, Japan.

Cautions for temperate climates include the potential for sensitive marine clays in Norway, Sweden, Finland, Russia, Canada, and Alaska. Sites with these quick clays should be considered unstable, and earthworks are not recommended. (See Chapters 6 and 9 for more details.)

In wetter regions, you will want to avoid the possibility of saturating a hillslope when there is a potential for landslides. In these cases, moderate installment of earthworks may be acceptably safe. For areas at risk of a slide, diversion of water and reforestation will help to avoid saturation and to stabilize the site.

The temperate regions of the world also tend to be the areas with the heaviest regulation. In many cases, open water storage will require special permitting, and the design and installation of a dam may require an engineer to plan and oversee the project, even for farm dams.

Design will also have to address the potential for decoupling landscapes, especially in drier regions. In some cases, this is legislated, and you will be allowed to build storage for only a certain percentage of annual precipitation that falls on your site.

Tropical climates

Tropical rainforests, tropical monsoon climates, and tropical savanna fall between the Tropics of Cancer and Capricorn, and make up approximately 35 percent of the Earth's surface. In these climates, particularly in rainforests, the influence of trees is more pronounced. Rain intercepted by the canopy of rainforests evaporates, humidifying the air, which increases local rainfall. This effect is carried throughout the forest, making trees a vital component of the hydrological cycle. Removing the trees will decrease precipitation.

In the wetter regions, evaporation will not be much of a consideration in open water storages. Evaporation will be a factor in monsoon tropics and tropical savanna, however. In these regions, the trend is toward greater aridity because of land use changes and climate change. As a consequence, you will need to address evaporation in open water storages. Deeper storages with less surface area help to curb evaporative losses. Strategies for shading water are also effective.

In the wet tropics, filling ponds and dams is not a challenge. Integrated support structures, such as swales, are usually not needed to fill dams, though swales or channels can be used to position spillways in desired locations. The major concern will likely be instability from saturation of the soil rather than maximization of infiltration. The wetter the conditions, the greater the risk of landslides will be. In the case of terraced beds for crop production, terraces should be limited to a maximum of eight benches in a set, with forest both above and below the terraces. The terraces will need to be drained to remove excess water to avoid soil saturation. On flatter land, such as in the footslope or toeslope, wet terraces can be more safely used for crops such as taro, rice, kangkong, watercress, water chestnuts, lotus root, and wapato (*Sagittaria latifolia*), among others.

Coral atolls

Coral atolls are delicate environments. The soil contains a calcite layer known as caliche, below which is the fresh groundwater on which the atoll's inhabitants rely. Any action taken on an atoll needs to take great care to avoid contaminating the groundwater supply. Harvesting water off of all available hard surfaces for storage in cisterns is advised.

The traditional approach to agriculture on atolls is to form a pit by breaking through the caliche layer. Ground water then supplies irrigation for taro production at the bottom of the pit. The sides of the pit can be terraced to support additional agriculture as well.

Drylands

Drylands make up 41.3 percent of all available land, and this figure is growing as more regions are dealing with encroachment of arid and semi-arid conditions. These regions suffer from decreasing agricultural yields, increasing population, and growing poverty. Without a careful water strategy, water wars, starvation, and environmental degradation loom on the horizon.

In drylands, evaporation is a major concern. Open water storages will be subject to huge evaporative losses unless shaded. Any open reservoir should be as deep as possible while maintaining bank stability and should have a minimal surface area. Shading open water storages, as described in Chapter 8, will help to reduce evaporation while providing a biomass or food crop yield. Any cisterns should be covered to help prevent evaporative volume loss.

A traditional Indian variation on dams is the *johad*. These are check dams built with the intention of capturing runoff to allow it to infiltrate into the soil and re-charge ground water. With a *johad*, no emphasis is put on creating a seal; nor is zoned construction used for the construction of the wall. A cutoff trench will help to create a more durable, long-lasting wall, but efforts to find suitable clay are not necessary. The point of a *johad* is to leak, or more specifically, to infiltrate water. Construction of the dam wall will use the same dimensions as the zoned construction highlighted in the "Dams" section of Chapter 6. The upstream bank is built with a 1:3 slope; the center is 3 meters (10 feet) wide; and the downstream slope has a 1:2 slope. Compaction is necessary for the integrity of the wall, but the wall itself is built of homogenous material, unlike a zoned-construction dam. If the bottom of the *johad* can be ripped without self-healing (lateritic soils that flow easily when wet are very common in India), this will improve infiltration rates, reducing evaporation and speeding groundwater recharge.

Infiltration is a very useful strategy in drylands, as soil moisture is less prone to evaporation than surface water. Ripping, swales (and the variations given in Chapter 7), land imprinters, *negarim* and *meskat* systems, and dryland terraces are used to infiltrate water. Due to the increased runoff volumes, interception features such as swales will have to be larger than they would be in areas where runoff is lower due to greater hydraulic conductivity of the soil and increased interception from vegetation. The diverting and spreading approaches used in spate irrigation direct and distribute water to agricultural systems, where water is infiltrated right where crops are growing.

Wherever enough water can be made to support them, trees should be planted and their growth encouraged. The benefits of trees include increasing local nutrient availability; providing food, fuel, timber, and biomass; increasing groundwater recharge; increasing slope stability; and decreasing erosion. From the deforestation of the headwaters in the Middle East 9,000 years ago to Australian deforestation for agriculture since World War II, the loss of trees accompanies decreases in rainfall and generally lower groundwater levels. Plant trees when possible.

Caution is needed when water harvesting in drylands. The risk of decoupling landscapes to the detriment of downstream communities is greatest in dry regions. Care in planning is needed to avoid overharvesting and to make efficient use of the water that is captured.

Lack of water is not the only hazard in arid and semi-arid areas. Rainfall tends to come in large, sporadic rain events, falling on soil that has higher runoff than other areas (with soils that are sometimes hydrophobic). This means that flash floods can arise in gullies and river beds, presenting a risk to people and property. Recall from Chapter 2 that the same water-harvesting systems that allowed the ancient city of Petra to thrive also put it at risk and eventually led to its downfall. Petra's location right in the wadi meant that when the dams failed and released the flood waters, a huge torrent washed through the city's core, doing tremendous damage. You will need to keep in mind how any earthworks you build will affect flow patterns and consider whether this will put lives or property at risk. You will also need to assume that your earthworks will fail catastrophically and ensure that failure will not pose a danger.

Some dryland areas contain diffuse clay, making dam construction very risky. Lime or gypsum can be used to help stabilize dispersive clays, but the consultation and guidance of an engineer is advisable.

Decoupling catchments

Be aware that installing earthworks can effectively decouple portions of land from the watershed. This is more of a concern in dryland areas with soils of low hydraulic connectivity. Being mindful of how much water we intercept in the head slope in these situations will reduce the impact your earthworks have on your downstream neighbors. Similarly, earthworks built in the base slope have a greater decoupling impact, with dams in the valley of base slopes having the greatest impact.

The path of water

Remember that runoff flows at 90° to contour. When you know the contour of the land, you know what areas will be wetter (head slopes) and which will be drier (nose slopes). Observation of the land will also tell you in what areas runoff is flowing. Bare earth will show evidence of erosion as water channels and flows downhill. In grassy areas, a large rainfall can sometimes flatten the grass as water flows overland. Hydraulic conductivity will also affect runoff volumes. Knowing where the different soil types are on your site will give you a better understanding of how water will behave on the site. The more carefully you observe the land, the better decisions you will make when it comes to designing earthworks for your site. Take your time in the observation stage.

Soil

Consult soil maps for your area to find out what soils are prevalent in your area. If your site has quick clay (see Chapters 6 and 9), recognize that hydrating the soil can potentially lead to liquefaction and landslide. Quick clays are very risky, and I have a policy of not developing earthworks on them. *Dispersive* clays (see Chapter 6) are also unsuitable as a structural component for dams. These clays essentially break apart when hydrated. Because of this, they are prone to failure if used in dam walls.

The greater the pore size of the soil on site, the more readily water will sink in, assuming that the soil is not hydrophobic. The need for earthworks, as well as their potential benefits, diminishes as hydraulic conductivity increases. There are, however, approaches you can use to build open water storages (shown in Chapter 6) in more porous soils.

Slope stability

Quick clays and dispersive clays are not the only potential hazard on a site. Slumping and other evidence of landslides are signs that further slides are a potential danger on the site. (See Chapter 9.) The general rule of thumb is to limit work to slopes of 20°, with terracing being an exception to this rule. Remember that the base material for soils (sand, silt, and clay) each have an angle of repose, which can change when saturated with water. Organic material in the soil as well as plant roots will increase stability, though this will change when you excavate a site. It is the tendency of all slopes to erode down to very gentle slopes eventually, given thousands or millions of years. You want to avoid making any extreme changes over the period that people will be inhabiting the site. In other words, avoid triggering slides.

You will also need to consider how much water you will need on your site. If the purpose is to have a reservoir for livestock, you will need less water than if you are irrigating crops or pasture with the water you capture. As you saw in Chapter 4, crop irrigation has a greater impact on streamflows than do livestock operations.

Finding contour

Even if you already possess a reliable topographical map of the site, you will need to mark out contour lines on the site for planning and implementation of earthworks. A map of the land will not tell you about protruding rocks or show any trees that you want to save. You'll have to assess the site for such obstacles yourself and work around them.

Laser level

There are a number of ways you can establish contour on a site. By far the easiest, fastest, and most accurate method is the laser level. Additionally, it requires only one person to operate. Laser levels are computerized versions of optical builder's levels that use an optical laser to establish a level line. Like contractor's levels, they are mounted to a tripod. The laser level has a tilt switch built in, which it uses to establish a level plane on which to send out its laser beam. As long as the tripod is relatively level, the laser level can find level on its own. Otherwise, an alarm will sound, indicating that the tripod needs a more level mounting.

As with an optical level, there is a staff with metric and imperial measurements printed on it. The difference is that a laser receiver mounts on the staff. Whereas an optical level requires two people to operate—one at the telescope and one holding the staff—a laser level requires only one person for operation. The receiver indicates whether the staff is above or below level, as well as indicating when level ground is detected. Most laser levels have both an LED screen and an audio alarm to indicate this. It is important to hold the staff vertically to avoid paraxial error, which means that if the staff is not vertical the receiver will be closer to the ground, giving you an inaccurate reading. To take care of this, many receivers have a spirit level to indicate when they are level.

The speed and accuracy of the laser level not only makes it the best tool for marking contour, it also makes it the idea choice for excavation. When you are building earthworks, you will regularly need to check contour. When you compare the time saved by a laser level versus the cost per hour of machinery, it is almost always cheaper to rent, or even buy, a laser level than it is to use other leveling tools.

Additionally, some bulldozers are equipped with laser receivers that can be used to automatically guide a dozer on a level path.

If there is a benchmark on the site (a point where the elevation is known), you can start at the benchmark, then mark either above or below that point. While it is nice to have a benchmark with a known elevation, it is not necessary. What is most important is relative elevation on a site. You might want to place earthworks at a certain elevation with respect to other objects on site, but rarely, if ever, will you want to set elevation with respect to sea level.

A laser level has a limited range in which the receiver can detect the laser. As with a dumpy (or builder's) level, you are limited by line of sight. If there is an obstruction, such as a person, building, tree, or ridge, between the laser and the receiver, you won't receive the signal. Likewise if the receiver is out of range. Move the tripod and laser to a position that can detect both your last established point and points further along the direction you are heading. Once you move it, the tripod will not be on the same level, so you will need to recalibrate the height of the receiver on the staff using a previously established contour point. Hold the staff at the known point and slide the receiver into position until it indicates level.

To find a higher or lower elevation from the elevation you are on, note how many feet and inches, or how many centimeters, the receiver is at on the staff, then reposition the receiver down on the staff by a set amount if you are going to a higher elevation, or up on the staff if you are going to a lower elevation. For instance, if the receiver is at 173 centimeters (approx. 5 feet, 8 inches) on the staff at the current contour, and you want to make another contour line 30 centimeters (approximately one foot) lower than where you currently are, simply reposition the receiver to 203 centimeters (6 feet, 8 inches) on the staff and establish a new line.

Ground is rarely smooth. Rather there are generally bumps and divots located across the surface. When establishing points, be sure you are not placing the staff on a mount or in a hole. This will give you an inaccurate picture of the overall landscape. Mark level points with stakes or landscape flags. Landscape flags come in a variety of colors. The different colors are intended to indicate different meanings in the field of construction, but you can use alternating colors to better mark out your contour lines. It is easier to view the contour lines if you use alternating colors as opposed to one color. This also eliminates the possibility of mistaking one contour line's flag for another and consequently digging off contour.

When you start to lay out contour, you will quickly notice that some points look off. It is extremely unlikely that this is the case. The human eye is very bad at looking at a landscape and detecting a level line across the surface. The only chance of error with a laser level is the receiver getting a false reading from a flashing light. It's not uncommon for equipment lights such as a backhoe's caution light to give a false reading. Even low flying aircraft have been known to create a false signal. Be mindful of this possibility.

Farmer's level

If you are concerned about the accuracy of your points, you can quickly check them with a hand-held farmer's level (AKA sight level). This device has a scope with cross hairs and a spirit level inside. You move to a position where you can see an established point is in the crosshairs when the bubble indicates level, then pan across the landscape on a level plane, checking your points. Farmer's levels also come in handy when you are doing a rough assessment of a site. When walking the land, you can use it to check contour and get a fairly accurate idea of what the land is doing. Keep in mind, however, that they are imperfect devices and are subject to error. They are intended for rough work, not precise measurements.

Builder's level

With a builder's level (also known as a dumpy level), two people do the work of the laser and laser receiver. One person remains at the tripod looking through the telescope, signaling to the other person holding the staff. Because of the time it takes to check the level in the telescope and signal to their partners whether the point is too low, too high, or correct, the builder's level takes considerably more time than a laser level.

Water level

The next best option, though quite a step down, is a water level. A water level consists of two graduated staffs (metric or imperial, depending on your preference), with a length of clear pex tubing filled with water attached to each staff. You start with an establishing point, placing a flag where the first staff rests. The other staff is placed right next to the first so that they are on level ground. You then read where the water level is against the measurements on the staff. The person with the second staff then walks along the hillslope to establish the next contour point, being careful not to spill any water. Each person then alternates walking across the slope, establishing points.

There are a few drawbacks to this approach. Firstly, it requires two people to operate. Secondly, it is easy to spill water out of the tube, making it necessary to recalibrate. Finally, the worst problem is that the plastic pex tubing softens in the sun. As it softens, the weight of the water causes the tube to bulge, which in turn causes the water level to drop. The bigger the diameter of the tubing, the less pronounced the problem is. Even with wider diameters like ⅝ inch, however, you can still visibly watch the water level drop in the tube on warm days. This will require recalibration with each point, making the process time consuming.

A-Frame

The final option is the A-frame level. The A-frame is the tool of last choice, a relic of a day before laser levels. It consists of a wooden frame in the shape of a capital letter A. From the apex of the level a plumb bob is hung so that the string extends lower than the horizontal member of the frame. A mark is placed on the cross piece where the string crosses, indicating level ground. The A-frame is calibrated on the most level surface available by marking the position of both legs of the A-frame, marking the position of the string, then exchanging the legs and marking the position of the string again. The level point is then half way between the two marks on the cross piece. Alternatively, if you find a level surface, you can fasten a spirit level to the cross piece. The legs must be level on the ground, but the cross piece is not likely to be level, meaning you will almost surely need to shim the spirit level before fastening it to the cross piece. This calibration must be done each time the level is used because wood expands and contracts with changes in humidity. If the level is not perfectly set, mark only every second point. The A-frame is "walked" back and forth from leg to leg across a hillslope, so errors are essentially self-correcting. Given leg A and leg B, if leg A establishes the first point, mark only contours using leg A.

When using the A-frame, you will need windless conditions, otherwise the wind will blow the plumb bob around, making it impossible to establish a level point. This is particularly an issue in deserts where winds pick up from sunrise and continue until sunset. In windy conditions, A-frames work only by using the spirit level option. The next problem is that it is possible to take points only where the legs touch the ground as the A-frame is walked across the land. The distance between the legs is constant, so if a leg lands on a mound or in a hole, that point will not be representative of the overall slope. Although an A-frame is a one-person tool, it is very time consuming to operate. Laser levels are obviously preferred.

LIDAR

LIDAR systems (short for Light Detecting and Ranging) are a fast and accurate means of generating maps. They work by sending a laser pulse across the landscape to be mapped. The reflected laser light is then recorded and used to provide mapping data. Like a laser level, it has a laser and a laser receiver, but it also contains a GPS tracking device. LIDAR systems are fairly accurate, comparing well with conventional topographic surveying but at a fraction of a cost. Drone-mounted units are becoming more and more common, and it is reasonable to expect that LIDAR will quickly become the standard approach to site mapping. With the data produced being digital, design work is facilitated, as the intermediate step of translating analog points to a computer is eliminated.

At the time of writing, I am working on a dam design on a LIDAR-mapped site. This particular project is an experimental one, involving Dr. Raul Ponce-Hernandez of Trent University in Peterborough, Ontario, Canada. In addition to having drone-mounted LIDAR for topographical mapping of the site, the drone was also fitted with an infra-red camera as part of an experiment in mapping soil types via infra-red. Should the experiment provide reliable and replicable data, this system will greatly facilitate site assessment.

As remarkable as LIDAR is, you will still need to establish contours on the ground with a level of some sort. The LIDAR makes the map, but you still have to translate the points to the real world for excavation. It can help you to build the model for designing, but it won't help you directly with implementation.

Mapping

LIDAR is by far the easiest mapping technique available. The next easiest approach is to establish contour lines on a site, then record their location with a GPS, marking waypoints at each flagged contour point as you walk a contour line. You will also need to record at least two other landmark positions that are clearly visible from aerial photographs. Driveway intersections and building corners often make good landmarks. Take waypoints at these landmarks. They will be your guide when overlaying your map of waypoints onto the aerial photo. In photo editing software, paste your digital map of waypoints in a layer above the aerial photo, scaling and rotating so that the waypoints for your landmark line up with their positions on the photo. The waypoints for your contour lines now indicate contour on the aerial photo.

Keep in mind that the accuracy of GPS devices is to within a radius of 3.5 meters (11.5 feet). Because of this, the margin of error is large enough that they should be used only on larger sites. Mapping an urban back yard with a GPS will be very inaccurate given the scale.

For most planning purposes, this approach is perfectly serviceable for planning the general location for earthworks. Remember that a map serves as a model for a site. The important part is that the intention is clear. The margin of error is small enough that it will not affect the actual implementation of the earthworks. As mathematician Eric Temple Bell pointed out, "The map is not the thing mapped." When you get on the ground, reality takes precedence over the plan. There is a chance that you will hit buried boulders or uncover unsuitable soil types that will require a change of plans.

Outside of LIDAR and the simple GPS method mentioned above, complicated mapping approaches using skilled surveyors with theodolites are required. The methods they use are too complicated to be described here.

Determining slope

As mentioned above, 20° is typically the steepest slope you will work on. To determine the slope of a hill, you can use one of several methods. You can either use an inclinometer, which is a device for measuring the incline of a slope; a theodolite, which is similar to a dumpy level, which can measure angles of inclination; or an inclinometer app or theodolite app for mobile devices. The mobile apps are the easiest method, and it is easier to download the app you need than to source an inclinometer or theodolite. Simply run the app and place your device on the backslope to determine the slope of the hill.

Earthmoving machines

For most projects, you will want to employ machinery to implement your plans. With all projects, there is the danger of falling victim to the planning fallacy. The planning fallacy states that when estimating the time required to complete a project, people overestimate their ability to get things done. An additional variation of this can occur with earthworks. Many people optimistically overestimate their ability to get work done by hand. A fit adult male can move about 5 m³ (6.5 cubic yards) of loose material, such as sand, per day and only 2 m³ (2.6 cy) of heavy material, such as clay, per day. By contrast, a small backhoe can move 53 m³ (69 cy) of loose material an hour and 35 m³ (46 cy) of heavy material, such as clay, per hour. In other words, earthmoving machines are at minimum 11 to 18 times faster than humans. Even if you are implementing projects in impoverished nations where the hourly wage for labor is very cheap, you can still do the job more quickly and cheaply with machines than by hand.

Clearly, earthmoving machinery is more economical in terms of cost and time needed to do a job, but is it sustainable? It may be. It is possible for an all human-built project to be unsustainable, so the same can be said of a machine-built job.

Unless the project serves no purpose whatsoever, this is unlikely to be the case, however. Looking at the sustainability of a project, the earthmoving machines carry embodied energy in the steel, plastic, rubber, and other components that go into the machine. Then there are the running costs in fuel and maintenance, topped off with the energy costs of the operator. The embodied costs are divided over the lifetime of the machine, and the fuel and operator costs are part of the immediate energy budget of the project. This is offset by the energy that the earthworks help to capture over their entire lifetime. Making water available helps to establish life on a site. Earthworks designed to sink water into the ground will support trees and other plants, which will become stores of energy for the environment. The establishment of plants builds a niche for more life to thrive on site. Similarly with open water storage, the aquatic environment will support more life with its intended irrigation purpose, but it will also support fish, stocked or naturally occurring, as well as other terrestrial and aquatic life the water attracts. The energy that the living systems capture becomes biological stores that offset the energy cost of implementing the project. These are just a fraction of the systems that can benefit from water harvesting, so it is easy to see that most projects will stand a very good chance of having a sustainable energy budget in the long run.

Types of machines

To implement your projects as efficiently as possible, you need to know what types of earthmoving machinery are available and what they are designed to do. We can sort machines into three types: machines that scrape and spread earth, machines that dig holes, and machines that carry loose earth. We will look at each type in turn. For information on estimating costs for these machines, please refer to Appendix 6.

Bulldozers

For many projects, it is likely that the first machine that you will use on site is a scraping machine, namely a bulldozer. Most bulldozers are crawler-mounted, meaning they are track-driven machines. While there are wheel-driven bulldozers, crawler-mounted dozers are more commonly employed as they have greater traction, allowing them to deliver more power to the blade, and they can operate over rough terrain. Wheel-driven dozers have a speed advantage and do not need a trailer to be delivered to a site. For the conditions experienced and the work needed on a water-harvesting project, a crawler-mounted bulldozer is usually your best choice.

Bulldozers come in a variety of sizes with varying blades and articulation of blades and may also have a ripper attached to the rear of the machine. Depending on the machine, bulldozers will be able to raise and lower the blade, tip the blade forward and back, angle the blade, and tilt the blade. A blade that will angle will push one end forward and pull the other end back, allowing material to be cast off to one side. Tilt allows one cutting edge of the blade to be lowered and the other raised, allowing the operator to cut into the earth at an angle. A *PAT blade*, short for power, angle, and tilt, can be a good choice for excavating swales. It will allow you to cut into the earth on the downhill side of a swale and cast the earth to the downhill side to build the mound. (See Chapter 7.) Similar to the PAT blade is the *VPAT blade*, which stands for variable pitch (the ability of the blade to roll forward or backward), angle, and tilt. There are three shapes of blades: S blades, U blades, and SU blades. *S blades* are straight blades, meaning the cutting edge along the bottom is straight. These are general-purpose blades used for clearing land and grading. They have the best depth of penetration out of all the blades. *U blades* have a "U" shape along the bottom cutting edge, with plates on their edges to help prevent pushed material from spilling out over the sides of the blade. *SU blades*, or semi U blades, are half way between an S blade and a U blade. PAT blades are often S blades with angling and tilting ability.

Some bulldozers have a laser guidance system that can be used in conjunction with a laser level. The dozer has a laser receiver that allows computer control of the blade depth. This automation greatly increases speed of operations, meaning fewer machine hours and lower costs. In the case of implementing swales, you would mark the contour for the operator to follow, allowing the laser guidance to set the blade depth.

Similarly, there are GPS-guided machines that contrast GPS data with a digitized site map such as a LIDAR-generated map. These systems allow machines to automate blade and directional control across contour on undulating landscapes. (It should be noted that LIDAR-generated maps have a vertical margin of error of ± 1.5 cm or 0.6 in.) The accuracy of these systems is quite good, allowing for increased efficiency. In the case of digging swales, you would traditionally mark a contour line on the site, then cut in the swale with the dozer, checking the depths with a laser level after the dozer has passed. Whether you would opt for this system depends on the scale of your project, machine availability, and costs. While it saves operating time and eliminates the need for establishing contour lines, it requires a digital site map and carries a higher machine operating cost per hour.

FIGURE 5.1.
A bulldozer with VPAT blade (with straight blade attached) and rippers attached on the rear.

Additionally, some newer dozers can even be equipped with remote control operation. Originally designed to operate in conditions where a human operator would be at risk, the external control of these dozers also allows for better observation of the blade.

Attached to the rear of some bulldozers is one or more ripping blades. The ripper can be used to loosen the ground to make the next cutting pass with the dozer easier. This feature can also be used to increase water infiltration. In the case of swale building, this is a very handy feature.

In addition to swale construction, bulldozers are the ideal machine for scraping aside and saving topsoil on large jobs such as ponds and dams. While there is an additional cost of machine delivery to a site, dozers can move topsoil far more rapidly than backhoes or excavators.

Excavators

If you have a job that involves digging, you will need an excavator. Excavators have a hydraulically activated boom, arm, and bucket. The boom and arm move the bucket into position, and the bucket does the digging. The larger the bucket, the more earth that can be moved at one time. The size of machine you use will depend on the scale of the job you are doing.

As with bulldozers blades, excavators have a variety of bucket types. The machine operator can choose the best bucket for the job, though for most jobs a

FIGURE 5.2.
Excavator.

trenching bucket is employed. One useful attachment is an angle-tilt bucket. This bucket allows not only the scooping motion used for digging (extension and flexion) but also a tilt from side to side so that the bucket can dig at an angle. In the case of digging swales, an angle-tilt bucket can excavate and groom swales quickly.

Loaders

Bulldozers scrape and grade. Excavators dig. When it comes to moving loose material, you will need a loader to move that material around. Consider excavation for a dam as an example. Some areas of the excavation site will have better quality clay than others. The excavator can make separate piles for high-grade and secondary-grade material, but the piles won't move themselves out of the way. To transport that excavated material requires a loader, such as a front end loader. While crawler-mounted loaders do exist, most loaders are typically wheeled machines, allowing for faster operation than continuous track machines. At first thought, a loader might not seem like an important machine to have on an excavating project, but it will save a lot of time on many projects.

Backhoes

Backhoes are smaller machines that have a rear-mounted excavator and a front-mounted loader. They are common machines, widely available almost anywhere in the world. Their hourly operating costs are quite low, making them the cheapest

FIGURE 5.3.
Wheel-driven front
end loader.

option in many cases. While backhoes cannot move as much material per hour as excavators, their lower per hour cost often makes them the preferred choice. If you are excavating a small pond, swales, or other light work, compare costs to see which machine will be the best option for you. (See Appendix 6.)

Compactors

In earthworks constructions, compactors are used to pack down soils to improve stability or to make a seal that will hold in water. Padded-drum rollers are used on larger projects such as dams or ponds. Many will come with a vibrating gear that vibrates the drum when engaged. This greatly improves compaction. These are often called sheepsfoot compactors, though a sheepsfoot roller technically has circular pads on the drum as opposed to oval or rectangular pads that are seen on padded-drum rollers. Sheepsfoot rollers actually compact most at the end of their pad, leaving the top layer less compressed. Because of this, padded drums are more desirable. Padded-drum rollers will have more traction in the uneven terrain of earthworks building and are therefore often preferable over smooth-drum rollers, particularly in wetter conditions.

In addition to rollers, there are also vibratory-plate compactors, which resemble a push lawnmower with no wheels. The compaction is achieved by a motor that

FIGURE 5.4.
A padded-drum compactor.

vibrates the plate up and down, though on a plane that is slightly pitched, allowing the machine to be pushed forward. There are also motorized rammers that are essentially motorized hand tampers. Both rammers and vibratory-plate compactors are suitable for small jobs only. Using one to try to compact a dam's keyway, for instance, would take a tremendous amount of time.

Safety

Machines are large, heavy, powerful, and above all dangerous. Large sites with multiple crew members will have more than one machine operating at a time. You must always be mindful of where the machines are. The hourly cost of operating large machinery is high, so you will want to give the machines the right of way. Don't hold up construction simply because you want to walk to a different area of the site. Also recognize that operators can't usually see what is behind them. Don't be in the way when a machine needs to back up. When working around something with a boom, such as a backhoe or an excavator, keep in mind that operators have a clear view directly in front of them where they are excavating. They have blind spots when turning to and from their dump pile, however. Don't be in the way when they are turning. It is also important not to exceed the machine manufacturer's guidelines for safe operation on slopes. Most machines are safe to use on cross slopes up to

15°, beyond which there is a danger of rollover. Injuries and fatalities occur most often when people ignore safe practices for the sake of time or out of over confidence. Always uphold safety standards.

Topsoil and erosion

There are a few good practices that should be observed on all sites. The first is recognizing the value of topsoil. Healthy plant growth relies on topsoil, and effort must be made to preserve it. I have experienced the horror of being asked to work on a site at which the site manager, a well-known landscape architect, scraped off the topsoil, built a large bund from the material, then topped everything off with subsoil from the site, bringing the excavating crew to the brink of tears. When I saw this, I refused the job. Scrape off the topsoil and set it aside to be redistributed on the land after the excavating and construction are complete.

When you have completed a project, you should mulch and seed the exposed earth to reduce erosion. Not only will banks erode easily without ground cover, they will also start to silt up what you have just completed building. On a larger project, hydromulching can be a quick solution to this issue.

References

Chang, Mingteh. *Forest Hydrology: An Introduction to Water and Forests*. Third Edition. Boca Raton: CRC Press, 2013.

FAO Corporate Document Repository. "VI. Aquatic Plants for Human Food." fao.org /docrep/003/X6862E/X6862E07.htm.

Mollison, Bill. *Permaculture: A Designers' Manual*. Tyalgum: Tagari Publications, 1988.

Rankka, Karin, Yvonne Andersson-Sköld, Carina Hultén, Rolf Larsson, Virginie Leroux, and Torleif Dahlin. *Quick clay in Sweden*. Report 65, Linköping: Swedish Geotechnical Institute, 2004.

Water Storage Techniques 6

Earthworks will involve digging a hole in the ground, mounding earth up above ground, or both. We can divide earthworks into essentially two categories: open water storage and earthworks that intercept or redirect overland flow. The distinction is not strictly delineated, with storage-type earthworks sometimes falling into the interception or redirection category as well. Open water storage consists of dams, ponds, and cisterns. The list for interception and redirection earthworks is long, and is addressed in Chapter 7.

Ponds

The simplest type of open water storage is the pond. A pond is simply a hole in the ground filled with water. It is differentiated from a dam in that there is no engineered retaining wall holding water in. In terms of safety, there is no wall to collapse. The only way a pond can catastrophically fail is if all the land around it is involved in a landslide. While this is something that might occur in regions of quick clay or other unstable conditions, it is not a normal occurrence.

The easiest and most inexpensive type of pond to construct is one that is excavated below the water table. Groundwater then flows into the pond, filling it. The deeper the water table, the lower the water level in the pond will be. Because they rely on ground water to fill them, such ponds are typically in the toeslope of landscapes. Depending on conditions, however, they can sometimes be located in the interfluve. This is more likely to be possible in clayey soils, where groundwater movement is not as swift as in more porous soils with greater hydraulic conductivity. If there is a spring on site, this area can be excavated for a spring-fed pond.

Similarly, if the site has a consistently waterlogged spot or an area that remains damp, it is a good candidate for a pond.

If the water table is too low, the pond will require some sort of a liner to hold water. Before you consider how you are going to seal the pond, you have to determine how it will be filled. In other words, is there enough catchment area to fill the pond? You determine this by calculating the volume of runoff that is available for the pond. If it exceeds the volume of the pond, then there is potential at that location. (See Appendix 2.)

Ponds can be lined in a number of ways. If your site has suitable clay, that can be used as a liner to hold water. (More on determining clay quality in the "Dams" section below.) Bentonite is another option. Bentonite is often available in large quantities as a livestock feed supplement at many farm supply businesses and is typically cheaper than bentonite packaged for sealing ponds. The volumes you need will depend on the soil on your site. If you have some clay on site, you will need less bentonite. If you are trying to build a bentonite pond seal on sand, you are typically going to need 30 centimeters (1 foot) of material. In such cases, bentonite would quickly become a major expense for the project.

Another lining option is a geosynthetic clay liner. This consists of bentonite that is essentially quilted into polymer geotextile. To use a geosynthetic liner, the bottom of the pond will need to be graded smooth. Once the liner is rolled out on the bottom of the pond, it is covered by a layer of earth. To help protect the liner, you can apply a layer of riprap (a layer of cobblestone) on top. This will help prevent accidental puncture of the liner.

Polymer liners can be used, but they are very expensive. If they are employed, the bottom will need to be groomed with a layer of sand underneath to protect the liner. As with the geosynthetic clay liner, earth and riprap should be placed on top to protect against accidental puncture.

The final option is to encourage the formation of soil colloids, known as gley or gleysol. To do this, you apply a nitrogen-rich organic matter, such as animal manure, or green manure, then bury it under a layer of sand. Gley forms in wet, anaerobic conditions, so if you are on a dry site, you will need to water the pond bottom and keep it moist to allow gley to form. Gley will be created as anaerobic bacteria break down the organic matter, forming a colloidal material. If conditions are right, this can create the seal needed. Gley is often used to seal leaky ponds (and dams) and is formed by penning livestock in the pond site. The combination of the animal manure and the compaction of the animals' hooves trampling the manure into the soil is often enough to fix a leaky pond.

Pond design

Once you have found a safe location with either sufficient catchment to fill the pond or groundwater near the surface, you can install a pond. The cheapest and easiest shape to design would be a simple geometric shape. Although this would be more economical, particularly in situations in which either a geosynthetic clay liner or a polymer liner is used, it misses out on an opportunity. By being creative with the perimeter of the pond, you can create more habitat for wildlife or stocked fish and open up more opportunities for aquatic-terrestrial interactions. Bays, points, drop offs, shoals and islands create structures that benefit aquatic life.

One approach to increasing structure in ponds and dams is to borrow from the Mesoamerican agricultural system known as *chinampas*. Though other cultures developed similar systems, this Aztec system has become the most famous. *Chinampas* were made by floating rectangular reed mats in shallow water and then piling soil on top to create a block for growing crops. In ponds and dams, a similar effect is achieved by excavating parallel fingers in the perimeter. As shallow water systems rapidly build soil, the bottom of the pond adjacent to the fingers can be periodically dredged and the soil added to the garden beds for fertility. The proximity to the water makes irrigation easier and creates a microclimate that moderates temperatures.

Pond construction

To start, scrape aside the topsoil where the excavation will take place and save it in a pile. Excavation will be done with an excavator or a backhoe, depending not he size of the job. The sides of the pond should be no steeper than 1:2 (26.57°). If you are installing a liner, you will need to prepare a graded bed of sand for the liner to rest on. A layer of riprap over the liner is recommended to protect it. You will need a permanent location for the excavated subsoil. At the end of excavation, return the topsoil to the portion of the project site that will be above the waterline.

Cisterns

Cisterns are set apart from ponds in that they are waterproof storage structures and are often indoor or subterranean. In hot desert climates, they are sometimes incorporated into a building to assist with cooling in addition to water storage. The ancient Nabateans, among others, used natural scour holes in rock, though a cistern can be any vessel that holds water.

Cisterns are the terminal storage for water harvested off hard surfaces, such as roads, rock faces, and roofs. Surface runoff is also an option, but a silt tank will be necessary to allow the sediment to fall out of collected water. (See the "Dams" section below and "Trees" in Chapter 7.)

A low-cost cistern originating in Libya is constructed via a brick-lined pit with two plastic liners. The first liner is permanent, whereas the second can be removed to make de-silting the cistern easier. The cistern is covered with a corrugated steel roof to reduce evaporation and to prohibit the formation of algal bloom. The runoff water from the roof of the cistern is directed into the cistern for storage.

Dams

Classification for dams varies from region to region, though the dams we will examine are all considered minor dams. The earthen dams—the type most employed by permaculturists—are under 6 meters (19.7 feet) in wall height and usually under 3,700 megaliters (3,000 acre-feet) in volume. Larger dams should be designed by an engineer and have their implementation inspected for safety.

Dam siting

There are a number of factors to consider when siting a dam. The first thing you should look for is a location where your dam can fail catastrophically without causing loss of life or property. If there are houses below the dam site, you will need to either build a suitable embankment that can redirect the water from a collapse safely or find a new location for the dam. While the construction guidelines below should provide you with a safe, stable dam, you should always operate on the assumption that the dam is going to fail. Caution at the start can help to avert disaster.

If the site you are considering is safe and has stable soils, you will need to ensure that there is enough catchment to fill the dam. In some locations, ground water might seep in from the bottom of a dam in much the same way it does in groundwater-fed ponds, though this is often something that occurs only in wet seasons. You cannot count on this as a means of filling a dam, however. The hydraulic conductivity of suitable clay is low enough that only minimal amounts of water would seep upward through the dam's clay seal. If you are in a location where a spring pops up on the landscape, or groundwater seeps upward and causes a wet spot on the soil, the soil can be unstable. Avoid these locations for dams. You will need to calculate the catchment area above your dam and multiply that by the volumes of runoff you can reasonably expect to experience on the site. (See Appendix 2.)

You will have to consider the needs you are fulfilling with the dam. Will the location of the dam conveniently serve those needs and be easily accessible for regular inspection and maintenance? Crop irrigation typically requires a larger dam because of greater water usage. Can you fit the volume of dam you need into the location you are considering? Is it possible to increase the elevation of the dam to maximize head pressure? This could eliminate the need for pumps. If the dam is for livestock, consider the ground that animals need to cover from dam to dam to obtain water.

With respect to hillslope, your dam will almost always be below the keypoint or backslope for two reasons. Firstly, above the keypoint you would be less likely to have the catchment needed to fill the dam. Secondly, on the steeper slope above the keypoint you would have to move more earth for less reservoir volume. You will also be working on slopes that are well within the safety limits of earthmoving equipment as you will usually not find sufficient quantities of clay on slopes greater than 8.5°. The toeslope is the zone of deposition for erased material. As a result, this is the section of a hillslope where the largest quantities of clay are deposited. Erosion carries clay particles downward from steeper slopes, depositing it in toeslopes.

You might also face legal restrictions in the amount of annual rainfall you are allowed to capture. The placement of the dam on the land will determine the amount of catchment you decouple from the watershed, which is an important consideration in dryland conditions. The closer to a stream or the lower in the landscape the dam sits, the more land that is decoupled. Similarly, placement of dams in headslopes or other points in a valley will decouple the land more than would placement on other points of the landscape. These sites happen to be the most economical locations for dams, however, as they require the minimum amount of earth moved per volume of storage. Also keep in mind that crop irrigation uses greater volumes of water than does livestock. Livestock dams tend to be fuller than dams for crop irrigation, meaning that more rainfall is surplus and can flow over the spillway and return to the watershed. In other words, they tend to draw less water from the watershed.

Appropriate soil

To build a clay-sealed dam, you need to know whether you have sufficient quantities of suitable clay on site. If your site has dispersive clays, quick clay, or highly expansive (AKA reactive) clay, dams are not possible, as these soils are just too dangerous. There are a few tests you can run on site to determine whether you have suitable clay.

Dispersive clays (AKA sodic clays) erode readily and are responsible for the gully-strewn landscapes seen in badlands. To test for dispersive clay at home, use the crumb test. For this test, cut a small cube of the site's clay, gently place it in a beaker of distilled water, and leave it for one hour. If the cube has either dissolved or has a ring of dissolved material around the base of the cube, it is a dispersive clay and is unsuitable for dam construction. This material is highly erosive and is subject to a phenomenon known as piping. Piping occurs when water erodes through the wall of a dam, forming a tunnel, or pipe, that widens as water passes through the tunnel. If you are uncertain of your clay quality, you can send samples to a soil lab to perform tests such as the pinhole test, in which distilled water is forced through a sample of material to determine how subject it is to piping.

Areas of Leda clay or quick clay (more in Chapter 9) are also unsuitable for dam construction. Leda is a type of newly formed marine clay (from 10,000 to 1.65 million years ago) that is mainly found in Canada, Alaska, Norway, Sweden, Finland, and Russia. Regions with "hard water" associated with limestone bedrock do not lend themselves to the formation of Leda clay, as the dissolved minerals from the limestone—mainly calcium and magnesium cations—replace the sodium and potassium cations that make Leda clay sensitive. Conversely, if the groundwater for an area is acidic "soft water," in which sodium cations are dominant, quick clay can form.

The binding together of clay and other soil particles is a process known as flocculation. As the sodium and potassium are leached from Leda clay, it begins to lose the ability to re-flocculate when the clay is disturbed. This means that when put under pressure, Leda clay can liquefy, leading to landslides on flat or nearly flat ground. Assuming you found Leda clay stable enough to use in the construction of a dam, natural flow of water through the dam wall (see the "Dam construction" section below) would begin a leaching process that would make the dam less and less stable over time. A disturbance on the dam wall could then trigger liquefaction of the clay and rapid collapse of the wall.

Expansive clay can also present a problem. Expansive clay, such as bentonite, is useful for sealing leaks in a dam or pond, but if a wall is constructed out of expansive clay, it can crack if it dries out. These cracks can then lead to tunneling and disintegration of the wall. This is actually less of a problem on larger dams than smaller ones, meaning that it becomes more important for the permaculturist building a farm dam. Similar examples would include basaltic clays, such as those found in Victoria, Australia, or in the Deccan Traps in India. The shrink–swell character of the clay can not only present a problem for the wall but also cause a shear failure in the soil underlying the dam, leading to a slump that collapses the hillside under the dam. The site will not only have to carry its own weight but also the weight of the water in the dam. This can be enough to reach the shear stress limit, leading to failure and slumping. These soils can be stabilized by the addition of hydrated lime, but the quantities required are very large (a volume from 3 percent to 8 percent of the total dry soil mass) and are unlikely to be cost effective. For these reasons, highly expansive clay is not a good candidate for dam construction. If you are uncertain about the clay on your site, a soil lab can determine the expansion index (a scale for categorizing how expansive a soil is). If the expansion index is greater than 91, the clay is highly expansive and an engineer should be consulted. It should be noted that all soils expand, and clays more than other soils. Expansion is a problem only in highly expansive clays, and it is reduced in compacted soils.

Assuming you have neither dispersive clay, quick clay, nor highly expansive clay, search for clay on the site by digging a test hole on your site somewhere along the point where your wall would be. Once you find clay and want to determine if it is suitable or not, there are some quick tests you can do. Silt and clay can be confused. Clay is sticky when wet and will coat your hands in mud that won't come off just by running water over your hands. You can take a sample of the material, moisten it, and roll it into a worm approximately the diameter of your little finger. If it cracks during the process and won't roll into a worm, it has too much silt to make a seal. If it does roll into a worm, grasp it in the middle and wiggle it. If it breaks apart, there is again too much silt. Join the ends of the worm to make a ring that will fit in the palm of your hand. It you can do this without the ring breaking, you have good-quality clay.

Another technique is to make a ball of the material the size of a tennis ball. Hold it at waist height, and drop it on a hard surface. If it deforms with no cracking, it is high-quality clay that will do the job of sealing your dam. If there is only slight cracking along the edges of the deformation, there is silt, but the clay should still do the job of sealing. If there are large cracks, or if pieces break off the ball, the material is too silty and will not make a sufficient seal.

Working with clay will give you a feel for the material fairly quickly. As you are overseeing construction, you will learn how to determine the quality of the clay just by grabbing a handful of material and working it in your hands. Squeeze the material together to make a small ball, then compress it with your thumb. Good clay is plastic and will mold around your thumb, forming a ribbon. If you are able to create a ribbon that is more than 5 centimeters (2 inches) long before it breaks, you are working with high-quality clay.

Assuming you find good clay on your site in a preliminary test hole, dig a minimum of four test holes along the proposed wall site at several points. Every 10 or

20 meters (33 to 66 feet) should suffice in this situation. Start at 20 meters, shortening the distance if you find a lot of inconsistencies. If you are in a valley, you will need to make more test holes to determine the nature of the soil across the span of the wall. You will also need test holes in the pit that you intend to excavate for the dam, though at less frequent intervals than for the wall. Test holes should be dug to 1 meter deeper than you propose digging. What you are looking for is quality clay and potential problems such as large boulders in the construction area.

Dam construction

For the construction of the site, we will be looking exclusively at the *zoned-construction* method. A zoned dam wall has a core of high-quality clay, an upstream zone of clay-rich impervious soil, and a downstream section of material with a greater hydraulic conductivity, allowing drainage. This method requires more care and time during construction than building a homogenous wall (in which the dam wall is constructed out of homogenous material) but is greatly preferred as it is more resilient and retains water better.

The parts of the dam include the dam wall and the borrow pit. The wall is divided into the key (Figure 6.1) and the uphill and downhill slopes. The *key* is the core of the dam wall and is filled with the best-quality clay from the site. The key sits in a trench called the *keyway* or *cutoff trench*. The upstream side of the wall maintains a slope no steeper than 1:3 (rise:run) and is built of good-quality clay to retain water. The downstream section of the dam is built at a slope no greater than 1:2, using freer draining material that is not compacted beyond just a bulldozer

FIGURE 6.1.

The basic elements of a zoned-construction dam.

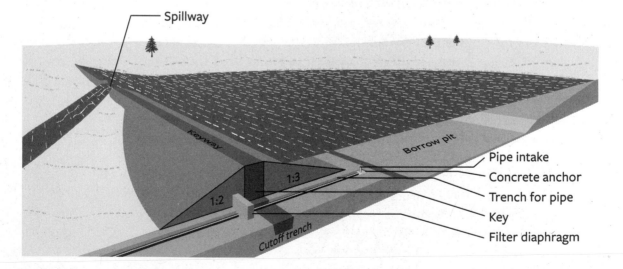

- Spillway
- keyway
- Borrow pit
- 1:3
- 1:2
- Cutoff trench
- Pipe intake
- Concrete anchor
- Trench for pipe
- Key
- Filter diaphragm

spreading the material. The maximum water level is maintained by a compacted spillway. Any water higher than the base of the spillway flows over the sill of the spillway and is safely discharged downslope. The height from the maximum water level to the crest of the dam wall is known as the *freeboard*. The height of the free-board is 1 meter minimum.

The dam will be built using the high-water mark as a baseline that will guide construction. Your dam is going to have a wall (two walls if you are building a saddle dam, described below) and a borrow pit from which you mine the materials to build the wall. This pit will add depth to your dam. Knowing where the wall will go, you can mark out its location with flags. Landscape marking paint is also helpful here. You will also need to mark out roughly where your borrow pit will be. Keep in mind that it will have to be placed such that its deepest point on the downstream side can be no closer to the upstream edge of the dam wall's crest than three times the height of the wall. In other words, the inside wall of the dam will have to maintain a 1:3 slope at maximum.

Scrape off and save the topsoil in a convenient location so that you expose the total area the dam will occupy. This means everywhere that will sit under water and all the area that will be worked on around the dam wall, which will be as wide as five times the dam's height plus the width of the core trench into which the wall will be built. Even if your topsoil is clay rich and appears that it would make a good seal for the dam, don't use it for this purpose. If compacted, the organic matter in topsoil breaks down over time, reducing the volume the soil would occupy. For this reason, it is not suitable as a material for wall construction.

You will excavate the keyway first. You will want to dig down so that you are 1 meter (39 inches) into the impervious clay on site. This will lock your key into the landscape's natural seal as well as creating a stable base for the wall. The key should be 3 meters (10 feet) wide at minimum, with bulldozer width being the standard (4 to 4.5 meters or 13 to 15 feet wide). If you hit bedrock, you will need a minimum of 1 meter (approximately 1 yard) of high-quality clay over the rock to ensure a good seal. To achieve this, you may need to punch through a portion of the rock to form a suitable cutoff trench.

As you excavate, you will separate out the highest-quality clay, which will be used to build the key. What is excavated from the keyway will not be sufficient in volume to form the key. Additional material will have to be mined from the borrow pit. Here, too, the material has to be graded in order of highest quality for the key-way, second best for the interior slope, and poorer, better-draining material for the downstream side of the wall.

Once the trench is excavated, compact the bottom with either a roller or a compaction hammer on an excavator. While a hammer is an option, compaction with a roller is by far the fastest approach and will likely be more economical. (See Appendix 6.) The keyway is then gradually filled in with the best-quality clay. Each load of material dumped into the keyway is referred to as a "lift," and must be put in no deeper than 15 centimeters (6 inches) at a time. Construction teams unfamiliar with dam building might argue for as much as 60-centimeter (2-foot) lifts. It is important to limit to lifts of 15 centimeters or less for proper compaction. As the material is placed in the keyway, it can either be smoothed with an excavator bucket or graded with a bulldozer. It is then compacted.

If the ground is too dry, the soil will not compact properly. In dry conditions, you will not be able to compress a handful of clay into a ball. You will need to water the site to the point where you are able to make clay start to hold together when you squeeze it.

In contrast, if the soil is too wet, the clay will become too plastic and will flow around the compactor rather than compressing. It can also lead to machines getting stuck. To test compaction, drop a shovel (handle-down) from a height of 30 centimeters (1 foot) above the ground. If the soil is compacting properly, it should leave only a faint mark in the clay. This test will also reveal when conditions are too wet. Even after multiple passes with a roller, the drop test will still reveal a distinct mark in the soil. In these conditions, you will need to cover the keyway and the reserved piles of clay and wait for drier conditions.

Continue the process of filling in the keyway with 15-centimeter (6-inch) lifts, compacting as you go. When you reach the top of the trench for the keyway, you will need to build up the upstream and downstream walls as you build the key. Compaction of the key itself is the most important part. Roller compaction of the upstream wall is not necessary. The crest of the wall is sloped inward slightly (around 20:1 slope or less) to direct any water that falls on the crest to be directed into the dam.

If you wish to have a pipe through the dam for irrigation, you will need to install the pipe as you are constructing the wall. The old method was to install anti-seep collars (AKA baffle plates or cutoff collars) along the pipe at equally spaced locations no closer than 3 meters (10 feet) apart but no farther than 14 times the distance that the collar projects from the edge of the pipe. In other words, an anti-seep collar that is a 60-centimeter (2-foot) square on a pipe with an outer diameter of 13 centimeters (5 inches), would have a maximum spacing of 6.76 meters (22 feet, 2 inches) between collars. The pipe is placed in a trench cut below the natural level of the earth deep enough to accommodate the pipe and the anti-seep collars. The aim with anti-seep collars is to try to prevent water from flowing along the pipe, which would lead to internal erosion and a partial collapse of the wall. As the wall is being built, a trench is cut into the natural "bank" earth perpendicular to the keyway. The pipe is installed and carefully compacted by hand or by mechanical tamper. Both ends of the pipe are fitted with poured concrete plugs where they exit the dam wall; this is to prevent vibration that could open space around the pipe and cause erosion.

This method is still in use, but since the 1980s it has been recommended less in favor of a filter diaphragm designed to stop internal erosion. The USDA's National Resources Conservation Service (NRCS) does, however, still accept anti-seep collars on low-hazard dams like our zoned-construction dams under 6 meters (19.7 feet) in height. Proper compaction of the pipe is hard to ensure with anti-seep collars, leading to many dam failures using this method. The NRCS requires filter diaphragms on all dams in which the product of the storage in acre-feet times the height of the dam from crest to downstream toe is over 3,000 acre-feet[2].

Failure with anti-seep collars occurs with internal erosion around cracks around the conduit, rather than water flowing directly adjacent to the outside edge of the pipe. Filter diaphragms address this issue of cracking by having a zone of graded sand around a conduit. Should an erosive flow start within cracks in the compacted clay, the particles will be carried to the edge of the filter zone, where it will form a plug at the interface of the filter. The filter is not intended for drainage. Rather it is to cause the formation of these plugs to seal internal cracks that occur due to uneven compaction around the conduit.

Filter diaphragm construction is actually quite simple and straightforward. (See Figure 6.2.) The pipe is laid in a trench dug into the natural "bank" earth perpendicular to the keyway, as described for the anti-seep collar method (i.e., deep enough to accommodate the pipe). Both the inlet and outlet are anchored into place with poured concrete blocks. The diaphragm itself is placed adjacent to the downstream edge of the key, aligned so that it is parallel to the wall. It will be 60 centimeters (2 feet) thick, but its other dimensions depend on the outside diameter of the conduit (D_o) you are running through the dam. The filter must extend under the conduit a minimum of 1.5 times the D_o. It must also extend 3 times the D_o on either side of the conduit, as well as above the conduit. The only exception here is if the maximum water level of the dam is less than 3 times the D_o, in which case it is limited to the maximum water level height. The filter must also always be no closer than 60 centimeters (2 feet) to the surface of the dam wall. As an example, a 6-inch pipe running through a dam would have a filter diaphragm that is 60 centimeters thick, 1 meter wide, and 84 centimeters (2 feet, 9 inches) high. The easiest method you can use to construct a filter diaphragm is to build the wall to the height you want to

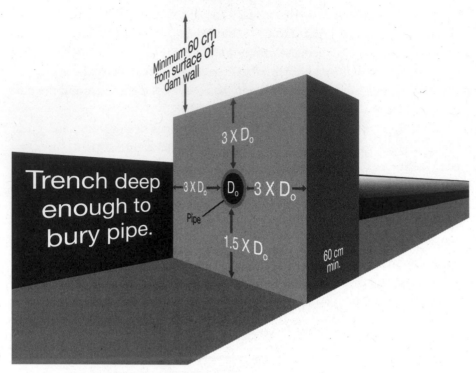

Minimum 60 cm from surface of dam wall

$3 \times D_o$

$3 \times D_o$ D_o $3 \times D_o$

Pipe

$1.5 \times D_o$

60 cm min.

Trench deep enough to bury pipe.

FIGURE 6.2.
The dimensions of a sand filter diaphragm. The diaphragm is placed just downstream of the keyway.

place the conduit, then cut a 60-centimeter (2-foot) wide trench that is 1.5 D_o deep, fill the trench with clean, well-graded sand (technically sand that meets the ASTM C33 standard), compacting it well as you fill the trench.

The inlet for the conduit needs a vertical riser to ensure that the intake is above the bottom of the dam. As clay expands and contracts, it is best to avoid completely draining a dam. The cracking that occurs when the bottom dries could possibly lead to a breach of the seal. The intake will require a screen to keep out fish and debris. An anti-vortex device—a vertical plate attached to the mouth of the inlet, dividing the inlet in half to prevent whirlpool formation—will help to maintain flow velocity in in the pipe when open. The outlet will need a to have a concrete or cobble splash apron to prevent erosion and should be clear of the dam wall.

As mentioned above, the maximum water level is controlled by a spillway. All excess water must flow over one or more spillways. Under no circumstances must you have water flowing over the dam wall. If this happens, you have an emergency situation and are at risk of the dam wall collapsing.

Spillways gently release excess water in a non-erosive flow. They are the dam's "safety valve" that protects the wall. They are designed so that their outlet is wider than the inlet to assist in smooth release of water. Table 6.1 has minimum guidelines for spillways for different catchment sizes, and Table 6.2 has the maximum water velocity guidelines for spillways based on soil type, set by the Food and Agriculture Organization (FAO) of the United Nations. Keep in mind that these are the bare minimum recommendations. It is a good idea to always overengineer safety features. Spillways are a very small part of a dam's overall budget. Extra investment pays itself back with a more resilient structure.

When constructing a spillway, make sure that the sill of the spillway is compacted hard and as close to perfectly level as possible. If one part is lower, flow will be concentrated there, leading to erosion and degradation of the spillway. The sides of the spillway must also be compacted and have a slope no greater than 1:2 (rise:run). The spillway marks the high-water point for the dam, but water levels will be higher than the spillway when water is flowing over it. There must be at least 30 centimeters (1 foot) of freeboard for periods of maximum water flow over the spillway. Widening the width of the spillway reduces the depth of the water flowing over it, reducing the erosive power the water has. As you can see in Table 6.2, the maximum water depths for spillways on more stable soil types can be quite high. You need to keep in mind the flow depth in relation to the freeboard height, as you cannot allow water to flow over the top of the dam wall. If you have water flowing over the spillway at 75 centimeters (29.5 inches) deep, you are dangerously close to

Table 6.1. Minimum spillway width recommendations

Catchment (Ha)	Spillway inlet width (m)	Spillway outlet width (m)
Under 20	3	7
20 to 40	6	12
Over 40	Engineer advised	Engineer advised

Table 6.2. Maximum spillway water velocities

Soil type	Max. velocity (m/s)	Depth at entrance (cm)	Discharge (m³/s)
Sand	0.30	15	0.05
Sandy loam	0.60	30	0.20
Sandy clay loam	0.75	50	0.35
Light clay	1.00	60	0.60
Heavy clay or gravel	1.25	75	1.00
Rock	1.50	150	2.50

Source: Stephens, 2010.

the top of the dam wall with a one-meter (39.4-inch) freeboard, and you would need to increase the height of the freeboard, meaning more construction costs, or lower the spillway, thereby reducing the dam's capacity. A wider spillway means a lower level over the spillway, lower construction costs, and a safer dam. The best course of action in terms of cost and safety is to install larger spillways than necessary. It is also good practice to install an auxiliary spillway. It is not unheard of for a spillway to become temporarily blocked in large rain events. Having more than one release for excess water is good insurance against such occurrences.

When placing a spillway adjacent to a dam wall, you have to ensure that all the water discharges cleanly away from the wall. You should not have released water making contact with the dam wall as it flows downhill. If possible, it is better to carry the water away in a level channel that leads to either a spillway at the end or to a swale with a spillway built into it at some point. The channel leading away from the dam will need to be fully compacted on all sides to carry the water away safely. Maintain the freeboard height for the downstream wall of the channel until the spillway. If you are joining the channel to a swale, do so only at a point where collapse of the channel would not endanger either the dam wall or anything downhill. After construction, seed the spillway with grass and mulch it lightly.

There should be a minimum of 30 centimeters (1 foot) of high-quality clay to line the bottom of the dam to seal it. A layer of sand or cobble riprap can help protect the seal and will serve as a guide when it comes to future dredging of the dam. Upon completion of the wall, redistribute the topsoil over the wall and on the bare earth that was cleared around the dam site. It is a good idea to mulch or hydromulch the site so that erosion and silting are reduced. If the orientation and length of the dam cause wave formation, you should have a layer of riprap covering the freeboard and extending down to the minimum depth of water on the wall. This will prevent wave erosion, protecting your wall.

Figure 6.3 shows a cutaway diagram of a zoned-construction dam and the phreatic line (water level traveling through the earth of a dam). Because of the highly compacted clay core in the key of a zoned dam, the phreatic line drops significantly as it passes through the key before continuing through the better-draining material on the downstream side of the dam. If the phreatic line escapes the soil's surface on the downslope side, it can make the saturated soil more plastic and prone to slumping and degradation. Other dam construction techniques avoid this by installing a drain of porous material near the center of the dam that drains out the downstream side or by installing a drain at the toe of the dam. The zoned-construction dam typically eliminates the need for drains.

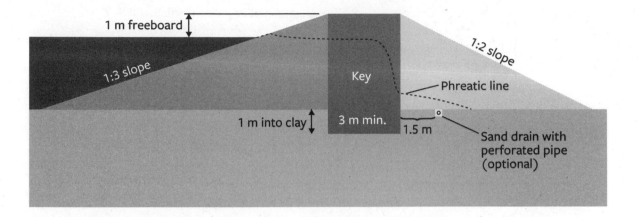

As an insurance measure, however, a 30 × 30 centimeter (1 foot × 1 foot) trench can be dug parallel to the cutoff trench on the downstream side, 1.5 meters (5 feet) from the trench. Add a 4-inch layer of free-draining sand, then a 4-inch perforated plastic drainage pipe wrapped in geotextile, and fill in the rest of the trench with sand. This will allow drainage of water passing through the dam wall and will help ensure that the phreatic line does not escape the surface on the downhill wall.

It is important to avoid creativity and stick to this time-tested construction technique. I heard of one dam in which the designers built a gently sloped swale on the downstream wall of the dam they constructed. You will want to avoid anything like this on dams you build. To encourage water infiltration on the dam wall, even inadvertently, risks raising the phreatic line and weakening the integrity of the wall. This particular dam also lies in a region of Leda clay, making the situation all the more precarious. Permaculturists like to make functional connections between elements. This is definitely a place where you do not want to do that. The dam wall has one job, which is to hold water in the dam. Trying to make it do more risks the integrity of the dam.

FIGURE 6.3.
Cross section of a zoned-construction dam wall. If the phreatic line reaches the downstream wall, it can weaken the wall.

Silt traps

Dams built in valleys can carry silt and sediment into the dam. In these instances, construct a silt trap above the point where sediment-laden runoff enters the dam. The aim is to provide a basin where the flow of water slows enough to allow the sediment to fall out of the water before allowing the runoff to continue into the dam. The passage from silt trap to dam can be made by either spillway or pipe, though pipes will carry less sediment into the dam and are the preferred method.

Note that the outlet end of the pipe will need a splash apron to prevent erosion. The outlet end of the pipe is made slightly higher than the inlet to slow the movement of the water as it passes into the dam. The silt trap will need to be inspected regularly and de-silted as needed. The silt trap is a stopgap measure for silting. You will need to encourage grasses above the dam to reduce silting.

The other situation in which silt traps might be needed is in dams that are fed by swales. Silt traps are dug in the swales at a point just before the water enters the dam. The water can then be carried to the dam via pipe or a level sill. Riprap at the entry points of swales will help reduce erosion of the dam.

Types of dams

The type of dam you build depends on the topography of the site. Most people think of *barrier dams* when they consider putting in a dam. These dams are built in a valley across a constantly flowing or ephemeral stream. These are the most expensive dams to build and maintain, require more careful engineering, and have the greatest impact on the environment. They can interrupt fish migration as well as decouple large portions of the landscape. Because they are built into a stream course, these dams will see more silting, which will require more maintenance to maintain the dam's volume. Having larger catchment areas, they require greater care in the design and placement of the spillway. There is also a greater risk in the event of a dam failure. In many jurisdictions, you will need special permitting to install barrier dams because of the impact they can have on the watershed. As the design and installation of these dams is far more complicated than the others presented here, it is advisable to have engineers design and oversee the installation of this type of dam.

Dams built in head slopes (AKA primary valleys) are an economical option for dams. The shape of the valley contains most of the water. This means that minimal wall construction is required to create the dam. The highest point in the landscape that these can be placed is such that the waterline is at the base of the backslope on the keypoint. These dams have been dubbed *keypoint dams* (Figure 6.4) after P.A. Yeoman's model of landscapes. Having a dam as high as possible in the landscape allows for greater irrigation options. If there is enough head pressure, pumps are not needed. There is also the possibility of flood-flow irrigation. (See Chapter 7.) The keypoint or the base of the backslope is the highest point these dams are built because to build higher would require moving greater volumes of earth for decreasing water-holding capacity. It is also more likely you will find suitable clay in the footslope than on the backslope. Walls in valley dams and keypoint dams are generally higher than for the other types listed below due to the fact that they have to be built up to match the contour of the high-water mark, plus 1 meter higher for the freeboard. This means you will need to find a relatively large supply of suitable clay to form the dam's key. In terms of safety, a collapse of the wall would lead to a concentrated flow of water, due to the placement in a valley.

It is also possible to build dams with the keyway following a contour line. The face of the hillslope itself can be straight, convex (ridge-like), or concave (valley-like). Known as hillside dams, or *contour dams*, these have the cutoff trench for the keyway excavated on contour, with the edges cutting back into the hillside to form the water-holding side of the dam. The waterline along the back of the dam is also on contour. The borrow pit is in the area inside the keyway, and its material is used for the wall. Whether convex, concave, or straight, contour dams require a lot of excavation and compaction to construct the dam, often making them a more expensive type of dam to build. These dams are sometimes placed in the shoulder of a hillslope, if there is sufficient catchment. This makes them valuable storages, as they are positioned high in the landscape. Availability of clay, and the economics of the earth-moved-to-volume ratio, will determine where these are placed. Contour dams are low impact in terms of decoupling land from the watershed.

A *ridge point dam* is a variety of contour dam. The dam wall is built on contour on a ridge, and the dam is cut out of the hillside. The cutoff trench for the keyway will need to be cut back into the hillslope, as with contour dams. The main difference with the ridge point dam is that a lot of material will have to be moved to construct the dam. In construction, you may find yourself needing to dump excess material somewhere else on site as you excavate the pit. The height of the wall might not be very high with respect to the natural grade of the land, especially when contrasted with a keypoint dam or other dams placed in valleys. Due to the flow of water over ridges, you might have an issue with catchment. You can use swales to direct water into ridge point dams when catchment is an issue. Ridge point dams decouple the least amount of the landscape from the watershed. They are also effective dams for wildlife as animals at the dam site have a larger field of view than they would in a valley.

In some landscapes you might find a suitable location for a dam that has one or more other low spots that would require a dam wall. In that instance, you can build a *saddle dam*. Building in a dip between two hill crests will require the greatest excavation of earth to construct the walls, as the walls will have to join the two hills. In other situations, you might find a valley that has two (or more, hypothetically) outlets that would require a wall to hold water.

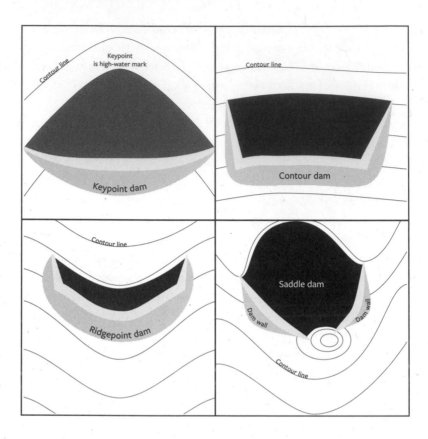

FIGURE 6.4.
Types of dams on
hillslopes.

Flat or nearly flat ground can also support a dam. In this situation, a *ring tank* can be built. Such a dam has one continuous wall that joins with itself, creating the reservoir. It can be a square or rectangle, but the most cost-effective shape—minimum perimeter to maximum area—is a circle. Material for the wall is mined inside and just adjacent to the wall. In an area of high rainfall, it might fill on its own, though it is typically filled by either pumping groundwater or diverting streamflow into the dam. With the borrow pit on the inside of the dam, ring tanks can maximize the volume. As part of the dam will be below grade, there will be a portion of the volume that is inactive, meaning it cannot be drained by gravity, necessitating pumps.

If the wall is built from material excavated from outside the wall, the dam is built above ground level, and is called a *turkey nest dam*. While the total volume is less with a turkey nest dam, the proportion of the active volume (the portion that can be drained by gravity) is greater than with a ring tank.

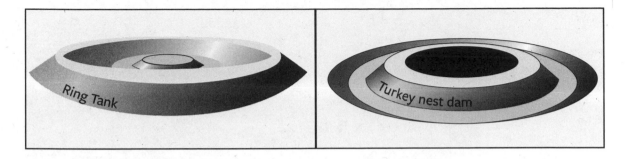

FIGURE 6.5.
Dams on flat ground.

Either a turkey nest or a ring tank can be built into the toeslope of a hillside. The portion of the dam abutting the hillslope is open, and swales are used to direct runoff into the dam. Being low in the landscape, this type of deployment of turkey nests or ring tanks can decouple greater areas of land from the watershed.

Check dams can be very effective in controlling erosion and encouraging groundwater recharge. Their construction is quite different from the earthen dams described above, as they are made of rock, concrete, or wood.

One of the simplest and fastest approaches to building check dams is to use gabions, which are galvanized steel mesh cages filled with rocks. In the case of erosion control, a cutoff trench is made in an erosion gully to anchor the bottom row of gabions in the ground. The trench must also extend to the sides of the gully, or water will flow around, eroding the sides of the dam. The empty cages are placed in the trench, then filled with rocks and wired shut. Additional cages are placed in a staggered fashion on top and wired to the cages below. The aim is for the gabion to back up sediment that would otherwise flow down the gully. The sediment will eventually reach the top row of gabions, so the downstream side will need a splash apron to protect the check dam. Though less stable than gabions, rock piles can also be used for check dams. In these cases, both the upstream and downstream slopes of the dam should be no steeper than 1:2, rise to run.

Start as high as you safely can in the gully and work downhill. The height of each dam added will be such that the crest is in line with the toe of the dam above it. In deserts, such check dams can fill after one or two rainfalls and will have water seeping out of the base of gabions for weeks after the rainfall. This opens a new opportunity for agriculture or re-greening efforts. Such systems are known as jessours (Figure 6.6).

The added expense of the steel cage can be a barrier in some instances. In these cases, rock walls can serve as a viable alternative. When compared to gabions, rock

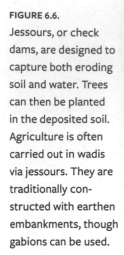

FIGURE 6.6.
Jessours, or check dams, are designed to capture both eroding soil and water. Trees can then be planted in the deposited soil. Agriculture is often carried out in wadis via jessours. They are traditionally constructed with earthen embankments, though gabions can be used.

walls require more care in the construction to make them stable. In drylands, there is often enough erosion that the inside of the wall will fill with soil after one or two rainfalls.

A *weir* is a check dam built into a stream bed. In most jurisdictions, the use of weirs in streams would require special permitting. In developing nations, such permitting is usually not needed. While this frees up possibilities, it is a privilege that must not be abused. Downstream communities relying on river flow are very often opposed to the installation of check dams above them. However, a properly designed weir will prolong the flow of water, not cut it off from those downstream. The aim of a weir is to slow water, not stop it. This is particularly valuable in drylands, where large rainfalls in a small number of rain events lead to sudden torrents of water inundating streams and rivers. There is a brief spike in flow, which drops off rapidly, or stops altogether. Prolonging the availability of water provides more opportunities to support communities and wildlife both locally and downstream.

As with erosion gullies, a cutaway trench that extends into the bank is needed to anchor the dam in place. Whether constructed of gabions, concrete, or log pilings driven into the stream bed, the center portion of the dam must form a lower spillway section to allow water to pass over the top. There must also be a splash apron on the downstream base of the dam. A gap or window in the wall or a pipe can be located near the window to ensure flow in all water level conditions and to allow fish to pass through when the water is lower. Concrete weirs are often built by creating a concrete trough along the river bed that extends up above the banks of the river. Large concrete blocks can then be lifted into the trough to form the check dam.

Troubleshooting and maintenance

All dams leak. Rapid drops in water level indicate a problem, however. If there is a point in the wall that has a leak, water will drop down to the level of the leak. If the dam drains right to the bottom, there is obviously a leak there. There may, however, be additional leaks above that point. This indicates the area where you need to concentrate on sealing the dam better. Drain an extra 30 centimeters out of the dam, then start work on sealing. You would attempt to seal first the wall above the waterline for a 1.5-meter (5-foot) band. When that portion is sealed, fill the dam and see if further leaks exist above this point. It is possible to have a dam with a persistent leak that resists efforts to seal it.

When patching leaks, mark out 1.5 × 1.5 meter grids on the wall (approximately 25 sq ft), adding 11.3 kg of sodium bentonite per 1.5 × 1.5 meter square (about 1 lb per sq ft), rake it into the soil, and compact it with a tamper. This approach is known as the mixed-blanket seal and requires less bentonite than the full blanket seal of pure bentonite. Sodium bentonite expands many times when wet, plugging cracks in the dam's seal. After you have raked in the bentonite, fill the dam to determine if it still leaks. If it still leaks at the same level, repeat the process on the back portion of the dam. If you have added bentonite around the complete circumference of the dam, and the dam still leaks, you could choose to repeat the process with greater volumes of bentonite or switch to other sealing approaches.

The same approaches listed above can be applied to sealing ponds, including those with gley formation and geosynthetic clay liners. There are also polymer sealing products that are becoming more popular. Water$ave PL is a polymer product made of polyacrylamide that has become very popular with dam builders in Australia. When a leak is present, it is applied to the water's surface as a dry powder at a rate of 100 grams per square meter. As it settles it forms a film that begins to expand to 200 to 300 times its original volume, sealing cracks. While it is a very popular product, some people are concerned about the use of polyacrylamide polymer and its theoretical potential to break down into acrylamide monomers. Despite testing, it is not clear whether the breakdown product, acrylamide, is harmful to humans or animals when ingested, except in larger doses than would occur through potential bioaccumulation from fish in a dam. In terms of safety, it should be noted that if you have eaten fried starchy food, in addition to many other natural foods, you have consumed acrylamide, as it is naturally occurring in many foods.

Additionally, polyacrylamide is regularly used as a soil treatment and erosion control. There is no evidence that breakdown of polyacrylamide occurs at a rate that makes acrylamide toxicity a concern.

The dam wall will need proper maintenance. Grass should be grown, but no trees must be allowed to grow on the dam. Clumping bamboo is a possibility, but no plants with a deep taproot or running roots, such as a running bamboo, should be grown due to the potential risk of piping or weakening of the wall.

Inspection of the wall when the dam is new should be carried out daily for a week when the dam is first filled. After one week, it should be inspected weekly for a month, then monthly for the remainder of the year. After that, inspections should be carried out annually at minimum. Record keeping is important with inspections. Take notes and photos of your findings. If LIDAR is available, having a record of LIDAR maps over time can alert you to a possible movement in the wall, which could signify a serious problem. LIDAR has a margin of error, but with multiple maps made over time, trends can be spotted.

Make measurements at set reference points on the wall. (Large stones can serve as reference points, as can steel marking stakes.) A pencil-thick steel rod can be used as a probe to test the downstream slope of the dam. If you detect softness, it suggests saturation of the soil, which could put the wall's integrity at risk. Look for a bow or a depression in the wall, erosive wear of the wall or cracks, adequacy of ground cover, visible leakage on the wall, animal burrows, or other anomalies. If you do find a problem spot, measure its location with respect to the reference markers, then make any needed repairs. The spillways will also need inspection, and maintenance may be needed. Inspect the area above the dam as well to see whether there are erosion issues that could contribute to silting. If there is a silt trap, it will need inspection and de-silting.

If there is a leakage on the wall, this is a potential problem spot. If you follow the guidelines and slope ratios for construction of the wall, seepage is unlikely to occur, however. If it is occurring, estimate the rate of seepage. Also note the clarity of the water. If the water is clear and in small, steady volumes, then material is not moving in an erosive manner through the wall. If the water is cloudy, coming through in large amounts, or is increasing in volume, this is strong evidence of piping or saturation that could lead to slumping. Here, the dam should be drained immediately until the problem is corrected. In this case, you should have an experienced engineer inspect and make recommendations. If there is excessive soil saturation and softness at the base of the wall, an engineer might need to be consulted about installing a toe drain at the base of the wall.

References

Agarwal, K.B., and D.K. Joshi. "Problems of earth dam construction in the Deccan Trap of India." *Bulletin of the International Association of Engineering Geology.* No. 20, 29-32 (1979) doi: 10.1007/BF02591239.

Federal Emergency Management Agency (FEMA). *Conduits through Embankment Dams Best Practices for Design, Construction, Problem Identification and Evaluation, Inspection, Maintenance, Renovation, and Repair (FEMA 484).* 2005. damsafety.org/sites/default /files/files/fema484/compressed.pdf.

Lewis, Barry. *Small Dams: Planning, Construction and Maintenance.* Leiden: CRC Press, 2014.

Mollison, Bill. *Permaculture: A Designers' Manual.* Tyalgum: Tagari Publications, 1988.

Rankka, Karin, Yvonne Andersson-Sköld, Carina Hultén, Rolf Larsson, Virginie Leroux, and Torleif Dahlin. *Quick clay in Sweden.* Report 65, Linköping: Swedish Geotechnical Institute, 2004.

Srithar, S.T. "Engineering design and earthworks aspects related to basaltic clays in Victoria." *Australian Geomechanics Journal* 49(2): 2014.

Stephens, Tim. *Manual on small earth dams: A guide to siting, design, and construction.* FAO Irrigation and Drainage Paper 64. Rome: FAO, 2010.

United States Department of Agriculture. Natural Resources Conservation Service. "Chapter 45 Filter Diaphragms," *Part 628 Dams National Engineering Handbook.* 2007. directives.sc.egov.usda.gov/openNonWebContent.aspx?content=17751.Wba.

United States Department of Agriculture. Natural Resources Conservation Service. "Pond. Code 378," *Conservation Practice Standard.* nrcs.usda.gov/Internet/PSE -DOCUMENTS/stelprdb1046898.pdf.

United States Department of Agriculture. Natural Resources Conservation Service. *Manuals, Title 210: Engineering National Engineering Manual 210-V-NEM (National Engineering Manual),* Part 520, Subpart C, DAMS. nrcs.usda.gov/Internet/FSE _DOCUMENTS/nrcs144p2_064805.pdf.

Virginia Department of Environmental Quality. Std & Spec 3.20. "Rock Check Dams." 1992. deq.virginia.gov/Portals/0/DEQ/Water/StormwaterManagement/Erosion _Sediment_Control_Handbook/Chapter%203%20-%203.20.pdf.

Westrup, Tilwin. Department of Agriculture and Food. "De-silting Dry Dams." 2016. agric.wa.gov.au/water-management/de-silting-dry-dams.

Yeomans, P.A. *The Challenge of Landscape: The Development and Practice of Keyline.* Sydney: Keyline Publishing PTY. Limited, 1958.

Yeomans, P.A. *Water for Every Farm.* Second Edition. Sydney: The K.G. Murray Publishing Company Pty. Ltd., 1968.

Interception Techniques

7

Swales

The usage of the word swale in permaculture is quite different from mainstream usage. In landscaping, a swale is a grassy drainage ditch. In permaculture, however, it refers to a level ditch designed to infiltrate water into the landscape. Outside of permaculture, this approach is referred to as a bund or a bank.

Swales are one of the most cost-effective water-harvesting techniques available. In contrast to pond or dam construction, the design, layout, and implementation of swales are quite affordable. Swales are a good fit on many sites, provided the site is suitable geologically and there is sufficient runoff to warrant their use.

Swales capture runoff, halting its flow and giving it a chance to sink into the ground. Recall that runoff will increase as hydraulic conductivity decreases. This means that swales are needed more on soils with smaller, more poorly connected pore spaces between soil particles.

When water enters a soil with small pore spaces, such as clay, capillary action will give water a stronger attraction to water than an adjacent layer of courser soil, such as sand. Counterintuitively, this means that water will move vertically and laterally to saturate a region of finer material before draining through an underlying layer of freer draining material, such as sand or gravel. After the soil is saturated, the gravitational force becomes dominant, drawing the water downward. Because good garden soil has friable soil with soil particles clumped together to form aggregates, creating greater pore spaces and greater drainage, swales above gardens typically don't do much to increase the water available for herbaceous plants. The water will saturate the soil with smaller pore spaces before moving to the soil of the garden. More friable soil is less likely to need swales as there will likely be less runoff. In dryland conditions where there is a lot of runoff, catching and storing water is almost always a good idea. Catching it above a garden is not going to hurt. In humid regions, the benefit is not as clear. Swales are better suited to groundwater recharge, tree propagation, and as a means of filling dams.

As mentioned in Chapter 4, runoff is more of an issue in drylands than humid regions, making swales much more beneficial and thus cost effective. Humid areas tend to see less runoff, particularly where there is ground cover. Often when swales are installed in wetter areas, they will fill with water, but the source of the water is primarily lateral subsurface throughflow and not runoff. There can still be a few instances in which runoff occurs. A rapid spring melt can saturate the soil, leading to sheeting overland. Additionally, very large rain events can lead to runoff. In both situations, the runoff can be intercepted by swales and redirected into the ground. Intercepting throughflow can be helpful when swales are intended to fill dams. In these cases, throughflow is directed toward the dam.

In terms of impact on the watershed, swales can decouple landscapes. However, Callow and Smitten (2009) reported that their modeling found no correlation between decoupling and length of swales when they examined decoupling due to dams incorporating swales in the Kent River basin in Western Australia. They found that placement in the landscape is more important. As with dams, the impact is greater lower in the landscape and in valleys. Swales behave differently from dams, however. Dams are intended to stop the flow of water and store it in the open. Additionally, dams are much more subject to evaporative losses than water in the vadose zone. Anecdotally, swales have been observed to increase the flow periods of streams or even lead to streams and springs appearing where they had been absent before. This would make sense as their primary function is to recharge the groundwater upon which both springs and streams depend, or to assist in tree growth, which is associated with increased rainfall and groundwater recharge.

Siting

Swales can be located on ground between 0° and 20° in slope, though the safe operating limit for most machines is 15°. The popularity of swales in permaculture leads to their being located in situations where they are unnecessary, counterproductive, or even dangerous, however. As mentioned, they are generally not effective support systems for annual crop production, though dryland conditions benefit from any help they can get.

Swales are sometimes mistakenly placed on hillsides in an effort to dry out waterlogged soil at the bottom of hills. If you install swales on a hillside, the footslope and toeslope will only get wetter. If there are buildings with basements toward the bottom of hills, you will need to estimate whether swale placement above them will create a flooding risk. Similarly, if there is a market garden in the toeslope, swales can result in waterlogging. In such a case, you would need to create raised beds, install drainage, or forgo swales in that location.

Another common mistake is to place swales in forested areas. As shown in the section "Trees" below, trees are already water-harvesting elements. Placing swales inside a forest is unnecessary, and cutting trees to make room for swales is wasteful. Forests and woodlots are already doing the job of water harvesting.

Finally, swales are not needed if there is no runoff. Gravels, coarse sand, and glacial till have zero or near zero runoff. Installing swales on these soils is a waste of energy and time.

The limit for swales is slopes of 20°, though most machinery is safe to operate only on slopes of up to 15°. Machine limits aside, the steeper the slope, the more you are subject to diminishing returns when it comes to the ratio of excavation to volume. Consider, for example, that you have a 1-meter-high swale on land that is sloped 3°. If the swale fills to the top with water, it will capture 954 liters of water per meter (76.8 US gallons per foot) of swale. If we are on a 20° slope, that same meter-high swale will only hold about 182 liters of water per meter (14.7 US gallons per foot) of swale.

In the case of flat ground, circular swales can be dug. Finding perfectly flat land is rare. Typically, you can mark out oval contour lines on ground that appears flat. In such cases, simply dig the swale with connecting ends, leaving you with a ring-shaped swale. Plants can then be established in the bottom of the swale's trench. Drylands would be the only place where such swales would be advantageous.

Many of the siting concerns faced with dams will apply to swales as well. Hydrating an unstable hillslope can lead to a slide if the soil is saturated. You will want to avoid hills with evidence of slumping or past slides. If there is a free face where earth has slid, you will absolutely want to avoid any swales above it.

Leda clay is also a problem. Hydrating Leda clay leeches out the salt that binds it, making it more unstable. Placing swales on Leda clay can prime a site for a landslide. Even soil that visibly appears flat will flow like water in a quick-clay slide. Avoid swales on Leda clay.

Another problem are would be in locations where saline seep is an issue. In such areas, sodium in the soil is dissolved when excess water enters the soil. The saline water then moves downhill with gravity and can move near the surface in low-lying areas where impermeable layers (which can also have high salt content) force the water table higher. The salt then accumulates on the soil surface, creating a saline seep. In these situations, swales are only likely to exacerbate the problem.

Orchards are another situation where swales might be a bad idea. Waterlogged soils can kill fruit trees. They typically need a minimum of 120 centimeters (4 feet) of freely drained soil to be healthy. Swales can increase soil saturation and raise the water table, which could put fruit trees at risk.

Placing a swale through an ephemeral stream bed will suddenly put a large volume of water into the swale. It will be unlikely that your swale will be large enough to hold the volume of water that would enter it, and the entry point will be a point of erosion and silting. There are situations in which large embankments are placed across ephemeral streams, but the banks are large, and require regular maintenance. (See "Spate irrigation" below.)

It is also possible that you might require a permit for placing swales on a site, or you might need to stay within a limit in terms of the amount of water you intercept. Some jurisdictions are against placing any alteration to the visible head of a catchment, and special permission will be required. There are also situations at high points in the landscape where you might install a swale around the top of a hillside to direct water to a dam, but in doing so you might be directing water from one catchment to another. In many places, this would be illegal.

If your site is suitable for swales, and you want to maximize groundwater recharge, then your swales will be more closely spaced at the top of a hill than toward the bottom. The reason for this is that the top of a hill is going to be the driest portion of the landscape. The two forces on groundwater are capillary forces and gravity. Water that enters the top of a hill will drain downward at a rate dependent on the hydraulic conductivity. Infiltrating water in the top portions of the hill will maximize its time on site while minimizing evaporative losses. As it travels underground, it will slowly add to the flow of groundwater that feeds rivers, springs, and wells. The simple way to lay out the spacing of swales to concentrate them toward the top of a hillslope is to use a logarithmic distribution.

On a given hillside, it is recommended that you limit the total number of swales to seven. This recommendation comes out of observation, not mathematical calculation. Appendix 4 gives you the formula for swales spacing and sizing.

Note that swales will silt up over time and will build soil as plants begin to grow in them. Left alone, they will eventually fill in and form terraces. If you are using swales as spillways for a dam, you will need to maintain them to ensure that excess water can flow through them.

Construction

When you have calculated the swales' location and size, you can mark the contours for the swales on site. With the contours established, you can begin excavation. Start with the top swale and work your way downhill, as the swales are calculated to take into account the catchment area above them, assuming other swales are present. Swales can be dug by hand, but given that it is on heavier soils with greater runoff that swales are needed, this is a job better suited to machinery. A backhoe, excavator, or a tilting angling bulldozer with an S blade or SU blade (also called a PAT blade or VPAT blade) can be used.

Irrespective of the method of excavation, the earth excavated from the trench is placed on the downhill side of the trench to form the swale's wall. This mound is groomed but not compacted so that infiltration is encouraged. Swale walls with a slope of 1:3 will reduce erosion, though they will need to be steeper in some situations (see Figure 7.1).

Excavation with a bulldozer can be a rapid way to cut in a swale. The downhill side of the blade is angled back and tilted downward. As the bulldozer cuts, it will cast the soil downhill to form the mound, which can later be groomed with hand tools or with a backhoe or excavator. If you have a laser-guided bulldozer, the process is very rapid. The contour lines are still flagged to give the operator the correct path to follow, but the blade depth is computer controlled, giving you a more accurate cut. For a larger swale, multiple passes with the bulldozer can be made to create the desired volume. When you reach the end of the swale, cut slightly uphill as the blade is gradually lifted to form the ends of the swale. Some bulldozers are equipped with a ripper on the back. If a ripper is present, ripping the bottom of the swale will assist in infiltration.

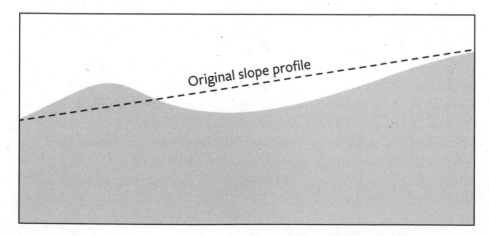

Original slope profile

FIGURE 7.1.
Cross section
of a swale.

Backhoes and excavators, while not as fast, are efficient tools for digging swales. Excavators will likely be the tool of choice for very large swales. Using these machines is a two person job. Set up a laser level with line-of-site so that the excavator does not obstruct the laser. The excavator crawls backward along the marked contour line, digging the trench and forming the mound. The other person operates the laser level staff, checking the level of the bottom of the trench as the excavator progresses. If you are not clear whether you should use a bulldozer or an excavator, an experienced machine operator can help you make the choice.

If you hit a large rock outcrop as you are digging, you can try to move it, work around it, or break the swale at that point and start up on the other side. You can determine whether the rock is bedrock or a boulder by excavating around it and trying to wiggle it. If it is secure, stand on one side (for your safety) and have the machine operator knock it with the machine. If you feel the ground bump like a small wave, it is a large but loose rock that can be moved. If the ground remains immobile, it's bedrock or too large to move.

If you cannot move it, you can try to dig around it either uphill or downhill. More earth will need to be moved in either case to keep the bottom of the swale level. Going downhill from the rock will require extra earth to build the wall up to a level point. Going above will produce excess material. The path to choose will depend on the shape of the rock.

Some designers suggest building swales slightly off contour toward ridges so that water congregates at the ridge to better hydrate these naturally drier areas of the landscape. This is a workable approach; however, digging gradually off contour like this takes much more time than just digging a level trench. Care needs to be taken to prevent rapid flow that would cause erosion. The process of readjusting the laser receiver at regular intervals can add significant time to the excavation process. The same effect can be achieved in two different ways without having to dig the entire swale on a gentle angle.

One method to encourage greater infiltration along ridges is to make the swale wider as it runs across the ridge. A wider swale effectively increases the volume it holds at that point. It also has a greater surface area with which to allow infiltration.

The other approach is to make the swale deeper at that point. The entire swale is level except for a deeper portion at the ridge. As water congregates in the swale, it will rise and flow to any low point first—in this case the ridge.

In both cases, you can increase infiltration with a broad fork or a spading fork. Just insert the tines into soil without turning the soil over or rocking the tool. For even better results, pour coarse sand into the holes and a thin layer over the entire surface of the swale along the ridge. The sand prevents smaller clay and silt particles from filling the holes and has a high hydraulic conductivity. This will allow rapid water infiltration and transfer of water to the soil below the surface.

Unlike a dam collapse, the collapse of a swale usually doesn't do much damage and is simple to fix with hand tools. If there is a point where water tunnels through the swale or spills over the top, it will erode in one place. You can make your swales more durable by installing a spillway, however. If you are starting near the summit, a large spillway will not be needed unless you have a large catchment with high runoff. Two- to three-meter spillways should be sufficient for subsequent swales. Table 6.1 can serve as a guideline when in doubt. Construct the spillway as described in the section, "Dams" (in Chapter 6). The outlet must be wider than the inlet. The side slopes of the spillway must be no steeper than 1:2; and the entire spillway requires careful compaction.

Spillways should be strategically placed. Don't place all the spillways for each swale so that overflow can cascade from spillway to spillway. Rather, stagger the spillways so that they do not line up as the water flows. If you can, place the spillways so that excess water will have to take the longest path to the next swale or off site into the catchment in the case of the terminal swale. Make sure that the overflow is not directed somewhere that excess erosion will be a problem or where flooding could create issues.

An additional benefit of a spillway is that less time is needed in grooming the swale. Without a spillway, effort is needed to maintain a level crest to the swale mound, as a low point would be a potential breach point where water could flow over the swale, cutting through the mound. With a spillway, you only need to ensure that all other portions of the wall are higher than the spillway.

When the excavation is complete and the swales are groomed, you can test the level of the swale bottom by observing it when it rains. Puddles will collect in low areas, while higher areas remain dry. After grooming, mulch and seed the bare earth or hydromulch it with a seed mix.

While swales serve as tree support systems, the swale mound is not usually an ideal place to plant trees. The mound is not compacted, meaning that trees growing on the mound are at greater risk of uprooting in strong winds. It's better to plant trees below the swale mound.

Case Study: The Green Tree Foundation swales project

At the invitation of Gangi Setty, founder of the Green Tree Foundation, I was part of a project to implement water-harvesting earthworks in Talupula, Andra Pradesh,

India. The project was more successful than predicted, and it serves as a good example of the potential that swales have for repairing damaged landscapes.

Technically, Andra Pradesh classifies as monsoon tropics. With climate change and the denuding of the landscape, the state is becoming semi-arid, with erratic and decreasing rainfall. The decreasing rainfall results in poorer agricultural harvests and greater abstraction and depletion of water tables as reliance on groundwater grows. The psychological effect of a browning landscape on the local population is quite devastating, and the hopelessness it engenders spills over to the economy, creating a vicious feedback loop.

The Green Tree Foundation, for its part, provides thousands of low- and no-cost trees for local farmers and residents in an effort to re-green the region. In addition to providing shade trees for roadways, they have supplied fruit-bearing trees for impoverished citizens, supported the local silk industry by suppling mulberry trees, and provided fruit- and fuel-producing trees for local farmers.

Before seeing the region, I had envisioned employing a variety of water-harvesting techniques. The conditions on the ground made swales the clear choice. Ripping (see the section "Ripping" below) the lateritic soils while dry—as it was when the project started—would have been extremely difficult and would have only chipped the earth into broken pieces. Wetting the soil would not help either, as the ground becomes quite plastic and self-seals. Building a dam would have required a reliable source of clay, which was not available. The lateritic soils are 8 meters (26 feet) deep or more in many places, and clay deposits are somewhat sporadic.

While swales were the obvious choice, we needed to know the volumes of rain that we might expect on site in order to size the swales correctly. Despite repeated requests over the years, the Green Tree Foundation was never able to obtain weather records from the government. As a result, we had to rely on local knowledge.

The site we selected belonged to Gangahadr, a local fruit farmer and long-time friend of Gangi Setty. The portion we were allotted for the project was a 7-acre section of hillside. While Gangahadr specializes in oranges, mangoes, and other fruits, on the hillslope he was able to grow only annual crops of pigeon peas because of the aridity of the site.

I started by mapping the site with GPS and observing the vegetation and erosion patterns to get an idea of how water was behaving on the site. After making some calculations, we designed the swale dimensions to be 4 meters (13 feet) across and 1 meter (3 feet, 4 inches) high to the spillway. The final plan would see four swales laid out on three contours. The runoff was estimated at 55 percent due to

FIGURE 7.2.
A section of the
project site. A group
of engineers are
using their dumpy
level to mark the
contours on site.

FIGURE 7.3.
A backhoe making
the initial excavation
for the swales.

the lack of vegetation and aridity. We had a crew of engineers mark the contours on the site with a dumpy level they had on hand.

We were able to hire a local backhoe, which dug 400 meters (1,312 feet) of swales with a volume of over 1 million liters (264,172 US gallons) in three days' time. We also had a crew of 10 local laborers to groom the site. As the soil is very hard in the dry season, they used steel pikes to break up the soil and groom the edges and mound of the swale.

The night before the final day of construction a pre-monsoon thunderstorm struck at 2 AM. Being very excited to see what would happen, Gangahadr rushed to the site and watched in the rain as the swales captured and infiltrated the runoff. The rain also made grooming the swales much easier.

In the weeks after I left, the Green Tree Foundation planted pioneering trees and expanded the mango orchard to the hillside. What was truly remarkable was that they were able to establish the mango saplings without the aid of drip irrigation, which is something that isn't normally tried, even on flat ground.

FIGURE 7.4. Workers groom the edges of the swale, using pikes to break through the hard soils.

FIGURE 7.5. A worker compacting a swale's spillway.

FIGURE 7.6. Tamarinds from an area of the site without swales. Photo by the Green Tree Foundation.

FIGURE 7.7. Tamarinds from a tree below the swales. Photo by the Green Tree Foundation.

The expansion of the mango orchard means that more workers will be needed to tend and harvest the fruit. This provides employment and improves the local economy. The trees themselves also assist in water harvesting and in nutrient accumulation. And there have also been higher water levels in a downhill well since the swales were installed. Considering the cost of the earthworks was under US$570, the project was extremely successful.

No project is perfect, however, and there are always lessons to be learned. The sides of the swale were steeper than I had planned. The laborers did a great job of smoothing out the swales, but they were unable to reduce the slope of the swale mound. Fortunately, this did not result in significant erosion. Additionally, although the mango trees were established without additional irrigation, their growth was slow. The addition of biochar (pulverized charcoal) as a soil amendment would have helped to establish soil organic matter. This would not only provide the trees with greater nutrients but would have helped with the soil's capacity to retain water. This approach was used traditionally in the Amazon and in Japan to assist in top soil formation and is a feature of many permaculture sites today.

Hügelswales

The permaculture affinity for combinations has led to the invention of the hügelswale. *Hügelkultur* is the European gardening approach of burying wood to create spongy, water-retaining beds for growing crops in. As a gardening approach, it has its upsides and downsides. It is a way of soil building and water storing. As the wood decays, it becomes soft and will absorb large quantities of water. Plant roots can burrow right into the decaying wood, tapping into it as a water reserve in dry conditions. The beds are often large enough that plants can be tended without stooping. They can be large enough that microclimates are created, with portions having full sun, partial sun, and shade.

The drawbacks are that often the wood could have been used for fuel or chipped for mulch. Mice often make nests in the beds, giving them easy access to crops. Wasps, too, often nest in the beds. And as the wood decomposes over time, the size of the bed will shrink.

Hügelswales are created by burying wood in the swale mound and using the mound as a garden bed. While it is space saving, this approach is not without the potential for disaster.

The problem with hügelswales is that wood—even decomposed wood—is far less dense than water. As water fills the swales, the buoyancy of the wood will create an upward force on the swale mound, weakening it. The weight of the water on the wall can then, at best, cause piping, gently blowing out the swale. At worst, it can float the swale mound and send wood and mud sliding down the hill, possibly to cause injury or damage property—a problem that has actually occurred with hügelswales.

Shrinkage of the wood overtime is another problem. Depending on the swale design, this shrinkage can create a low spot that can lead to a breach of the swale mound.

There is little risk if you are building a small hügelswale with little volume. In that case, a blowout of the mound will only create a nuisance. The risk comes in a larger wall. The only way to make it secure would be to hammer rebar into the ground angled uphill at a 45° angle to hold the wood in. The problem here is that the use of hundreds of pounds of steel for the sake of a garden bed that could simply be placed in a safer location is not sustainable. Additionally, swales are not particularly effective garden irrigation strategies, as mentioned above. If you insist on hügelswales, keep them small and on a small catchment with minimal runoff.

Contour bunds

In drylands, a variation on the swale can be used to collect and hold water for trees in a similar manner to the *negarim* technique (shown below in the section "Trees"), taking advantage of the higher rates of runoff in dryland areas. Swales are cut across land on contour but at a closer interval than the swales mentioned above. Bund spacing is from between 5 to 10 meters (16 to 33 feet), with mounds from 15 to 65 centimeters (6 to 25 inches) high, depending on the slope, using the same figures for the *negarim* system.

The difference between these contour bunds and swales is that wings are built out perpendicularly from the mound of the bund and a depression is dug out in the trench, adjacent to where each tree will be planted. The perpendicular wings are spaced 2 to 5 meters (6.6 to 16.4 feet) apart along the bund and extend uphill for at least 2 meters. The depression is dug to 40 centimeters (16 inches) deep and is placed near the corner of the bund and the wing. The tree is then planted between the depression and the wing as shown in Figure 7.8.

Semi-circular and trapezoidal bunds

Also similar to swales are semi-circular and trapezoidal bunds. As with the contour bunds described above, semi-circular bunds are crescent-shaped mounds that are used to capture water on slopes. Tree planting is then done at the apex of the crescent. This approach is also useful in establishing herbaceous plants. Trapezoidal bunds consist of three joined mounds that trap flowing water. Both styles will be made more durable if the ends are capped with riprap to reduce erosion in the event of overflow.

Larger versions of these bunds can hold fields that back up with runoff water, allowing for agriculture and groundwater recharge. Bund construction is more akin to a dam wall than a swale, though not a zoned construction as described in the

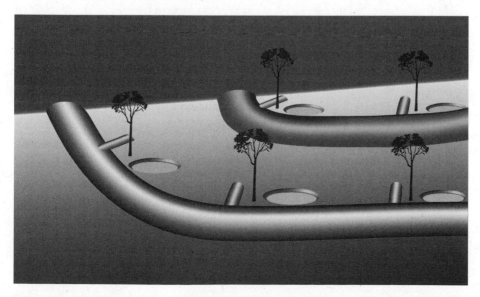

FIGURE 7.8.
Contour bunds.

section "Dams" (Chapter 6). Rather, the bunds are mounded earthen walls with neither a key nor a cutoff trench, though wall dimensions are the same, to make the bund more durable (i.e., 1:3 upstream slope, a core 3 to 4 meters (10 to 13 feet) wide, and 1:2 downstream slope). The spillways are typically built on either end of the retaining bund. If sufficient regular runoff escapes the spillway, additional bunds are built downstream. In permaculture, such a system, with the bunds completely enclosing a field or series of fields, is referred to as a "limonia" system when it captures water off a large rock such as a rock inselberg. Outside of permaculture, they are referred to as a *meskat* system, irrespective of the catchment surface.

Bench terraces

Terracing is a strategy for managing water and erosion that has been in use for at least 4,000 years. It can be used to slow or stop the flow of water downhill, reduce erosion, and provide a space to grow crops. Though many terracing techniques are more akin to swales and contour bunds, the bench terrace is the iconic method most people associate with the technique. As with swales, terraces are built on contour. Most every continent has a long history of using terraced agriculture for cropping grains, vegetables, and orchards.

Bench terracing is typically employed on slopes from 7° to 33°. It is possible to terrace very steep land of 45° or greater, though terracing such slopes takes tremendous effort. As with dams and swales, careful consideration of the site is necessary. In areas of unstable soil such as quick clays, or where there is evidence of slumping or sliding, terraces might carry with them the potential to trigger a slide. In areas of high rainfall, proper drainage will be necessary. It is possible to saturate a hillside and essentially "float" it, triggering a slide. At saturation, the angle of repose for a given soil is decreased, making it more prone to sliding. Soils are also going to have to be deep enough to build a bench. To limit erosion, it is recommended that terraces be a maximum of 100 meters (328 feet) long.

The style and construction of a bench terrace will depend both on your site and the on intended use of the terrace. Bench terraces can be either *continuous*—meaning one terrace right next to another—or *discontinuous*—meaning that there is a stretch of unaltered hillslope between terraces.

While either type can be used regardless of climate, typically discontinuous terraces are used in drylands. The portion of unaltered hillslope above each terrace serves as a collection point for runoff, which falls on the terrace, providing water for the crops planted on the terrace. In very wet areas, such an additional collection area for runoff would require additional drainage capacity for the terrace.

In addition to continuous and discontinuous terraces, the bench itself can be reverse sloped (sloped inwards toward the hill), level sloped, or outward sloped. *Reversed-sloped benches* are recommended in wet regions. They are sloped inward at a 5 percent (2.86°) slope. The reversed slope prevents water from spilling over the bench and down the riser, which, in conditions of heavy rainfall, could erode the riser, leading to a collapse. The excess water is collected in a drain in the inside edge of the terrace and safely removed. Drainage is handled by stone-lined collection drains that carry the water downhill. *Level-sloped benches* are typically associated with crops that need a level surface for flood irrigation, such as rice. *Outward-sloped benches* are used in drylands and have an outward slope of 3 percent (1.72°). The outward-sloped bench is intended to allow the spread of downward flowing water across the bench to irrigate the crops (see Figure 7.9). A retaining bund is recommended for each type of terrace to prevent excess water from spilling over the bench and down the riser.

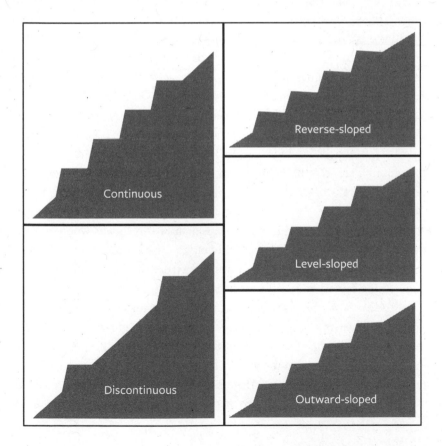

FIGURE 7.9.
Types of benched terraces.

Construction

When building by hand, terraces can be built from the top down or bottom up, though the latter is generally more practical. To build the riser for the bench, earth is cut from the uphill side and used as fill on the downhill side. If you work from the top down, you will be burying the best topsoil in the base of the riser. If you work from the bottom up, you can form the bench, then clear the topsoil from the section of hillside above the current bench, and spread it on the top of the bench. If you use a bulldozer to form the benches, you will run the dozer up and down the slope, pushing the dirt downhill to build up the riser. Appendix 5 has formulas to design benches with the minimum amount of earth moved.

The width of the terrace depends on the slope of the hill and the depth of the soils, with steeper slopes and shallower soils requiring narrower terraces. The intended use also determines how wide the terrace should be. For cropping herbaceous plants, the terraces will need to be wide enough to accommodate beds and pathways from which people can work the beds. For trees, the terrace can be under 2 meters (6.5 feet) wide, and may only be 1 meter wide. For an orchard, the terrace needn't run across the entire contour, unless erosion control is also an aim. Rather, the terrace might only be 1 or 2 meters (3.3 or 6.6 feet) in length. Furthermore, it needn't have the rectangular profile of regular terraces. It may instead have a circular shape along both the cut and fill of the bench.

As you build the terraces, you will need to compact the riser in order to build a stable platform for the bench. Stone retaining walls are not a necessity, but they will help to make the terrace more secure. Even a riprap cover on the riser will help. When building a stone wall, the first row should be buried to $\frac{1}{10}$ the height of the wall. For example, a 1-meter-high (3 foot, 3 inch) wall will need the base of the stone wall buried a minimum of 10 centimeters (4 inches) deep. Though a retaining wall is not necessary, if you are building one, inserting rot-resistant wooden ties into the riser at regularly spaced intervals will help to create more friction, which will reduce shear forces, as well as resist outward pressure that could collapse the wall. Attached to the buried end of the tie is a second horizontal "deadman" tie. The weight of the earth on the ties supports the wall. Charring the outside of the wood will greatly increase its resistance to decay. Sloping the riser inward toward the hill will make the wall stronger. The more gentle the slope, the more stable the wall will be. However, the more angled the wall is, the smaller the growing area. As soon as a riser is built, it must be planted with grasses. The ideal method is hydromulching, but it must, in any event, be planted so that the riser has protection against erosion.

With both level-bench terraces and inward-sloped benches, the bench will need protection against any water running down from above onto the bench. Placing a layer of stones against the bottom edge of the next terrace's riser will serve to protect against any erosive flow from above.

Ripping

Ripping is achieved via a chisel plow (AKA subsoiler). Chisel plows are often employed in no-till and low-till farming as a means of loosening and aerating soil. Patterned correctly, ripping can be a powerful water-harvesting tool, however. It can intercept runoff water and assist in infiltration. Whereas swales achieve this by having trenches intercept water, ripping achieves the same end by cutting narrow furrows in the ground. Any runoff is intercepted and enters the furrows to be absorbed into the soil. The furrows not only make water more available to the soil, they also allow greater oxygen exchange in the soil, both of which support soil life and assist in the natural process of humus production.

In the 1950s, Australian mining engineer P.A. Yeomans developed a model for understanding landscapes (see Chapter 3) and making water available for those landscapes, which he coined the *keyline system*. The keyline system was the model upon which permaculture founder Bill Mollison developed permaculture's water-harvesting approaches. Yeomans' method for ripping land was an integral element in his keyline system.

Chisel plows consist of shanks with pointed shoes at the bottom that cut through the earth, leaving a small furrow in the wake. In addition to the furrowing, the shoe can also help to loosen up compacted soil, essentially fracturing it and opening up fissures in the soil. The furrow allows easier water infiltration and improves soil aeration. This makes ripping a valuable water-harvesting technique as well as an aid to plant growth and improving soil health.

There are a number of chisel plow manufacturers. Most modern designs have straight shanks and replaceable shoes. Many are equipped with a coulter in front of the shank to allow it to cut through sod more cleanly. Some have injectors that allow the addition of liquid fertilizer. The Yeomans family is a manufacturer of high-quality subsoilers, some with very convenient features, such as a system that allows the shanks to pop out of the ground if they encounter a rock. In terms of tractor power needed, 15 horsepower per tine is generally recommended.

Ripping is not for all conditions. There will be little to no benefit on soils that are essentially self-healing. The lateritic soils that dominate most of India are a

FIGURE 7.10.
Chisel plow.

good example of an unsuitable soil. During the dry season, the soil is indistinguishable from concrete and can be broken only with picks or heavy equipment. Assuming you could manage to break the soil and pull a subsoiler, it would leave in its wake broken earth resembling pavement shattered by a jackhammer. When lateritic soil gets wet during the monsoon season, however, the soil becomes quite plastic. Walking on it when wet resembles walking on a mattress, and any ripping done would self-seal. Similarly, sands with no organic content, such as in an abandoned gravel pit, or coarse glacial gravel tills would simply self-seal when ripped.

On soils that are somewhat prone to resealing, ripping too deep can lead to furrows filling in after rains. If you are on loose soils with low humus content, check after rains to see if the ripping closed up. If it has, make an additional pass with the subsoiler. The injection of a slurry of liquid manure and biochar will help these soils with low humus content.

The approach that P.A. Yeomans developed revolved around the keypoint at the base of the backslope in head slopes (what Yeomans called *primary valleys*) that falls off a larger main ridge in the landscape. He followed the contour of the keypoint along the head slope and side slopes, coining the term keyline to describe that contour. This then became the guide for ripping with the subsoiler. The contour of the keyline is followed with the tractor, ripping the keyline. All ripping above the keyline (within the safe operating limits of the machine) and all ripping below it is done parallel to the keyline. Once the end of the side slope is reached, plowing is done slightly downhill at a 500:1 drop toward the nose slope (what Yeomans coined the *primary ridge*).

This pattern can be followed across the landscape from head slope to nose slope to head slope, and so on. It should be noted that the keyline from one head slope is likely to be on a slightly different contour than its neighboring keylines in adjacent head slopes. Typically the keylines become higher as the main ridge gains in elevation.

The effect of this cultivation is to coax the water that is flowing toward the center of the valley to spread more evenly across the land. The off-contour plowing from the end of the side slope moving downhill toward the ridge of the nose slope is to help to direct water toward the ridge line. Ridges are the driest portion of the hillside. This pattern helps to direct water toward the ridges to hydrate them better.

As this pattern is extended in parallel above and below the keyline, the ripping will be slightly off contour so that it directs water away from the valley and toward the ridges, creating a more even distribution of water across the land. When working on the summit or the toeslope, where there is no significant valley or ridge formation, the ripping is done on contour. If there is a valley shape without a keypoint on the site, you can use a contour high in the landscape in the valley and plow in parallel passes from that contour.

Yeomans recommended different approaches for cropland and pastureland. For cropland, he recommended one pass at 7.5 to 11.5 centimeters (3 to 4.5 inches) with the tines at 30 centimeters (12 inches) apart, followed right after by an 18- to 20-centimeter (7- to 8-inch) pass. The following year, a pass is made with weed knives (horizontal knives attached to the shank to cut the roots of weeds) set at 7.5 centimeters (3 inches) deep. Each year the plow goes slightly deeper, until it is at maximum depth, then moved up to 23 centimeters (9 inches) deep with weed knives set at 15 centimeters (6 inches). One benefit of the weed knives is that they leave organic matter in the soil, which could decay and contribute to humus production.

The approach for pastures is slightly different. Instead of two passes in the first year, only one pass is made at 11.5 centimeters (4.5 inches), with the tines spaced 30 centimeters (12 inches) apart. If resealing occurs during this first year, make a second pass at the same depth with 60-centimeter (24-inch) tine spacing. In the second year, another pass is made at 13 to 18 centimeters deep (5 to 7 inches) with the tines 60 centimeters apart (24 inches). In the third year, the plow is set to 18 to 25 centimeters (7 to 10 inches) deep with the tines 90 centimeters (36 inches) apart. Spring or possibly late fall are more ideal times for ripping, as it is best to keep cattle off the ripped area for a few weeks after the first rain post-cultivation.

Ripping can also be helpful for flooded land. Aeration from the plow will help to break the anaerobic cycle that waterlogging sets up. This will improve pasture quality.

It should be noted that ripping will increase infiltration, which will, in turn, increase throughflow. This will mean less runoff on the land. If a swale or diversion drain is placed below ripping, it can fill with throughflow, which might then be directed to open water storage.

Land imprinters

Invented in 1976 by Dr. Robert Dixon, the land imprinter is a tool for reestablishing grasslands in arid and semi-arid environments. The tool consists of a large drum that is pulled behind a tractor. The surface of the drum has offset angular teeth that leave a V-shaped divot in the ground as the machine passes.

The teeth on the imprinter can break through hardpan crust on the surface that would otherwise resist water infiltration, though ripping prior to imprinting may be necessary on compacted soils. The shape of the imprint will funnel both water and organic matter to the bottom of the divot. Dryland soils typically experience high levels of runoff. After an imprinter makes a pass on the land, the soil has the capacity to capture rainwater and allow it to infiltrate, rather than allowing the water to sheet off the land. The increased infiltration and the shape of the divot both serve to decrease evaporation. Leading the imprinting drum is a seeder that distributes a seed mix on the ground that the imprinter then pushes into the soil, where it will wait for the next rainfall.

The imprinting teeth on the drum must be under 25 centimeters (10 inches) long, measured along the imprinting edge, in order to discourage the formation of rills as water moves into and along the divot. The teeth typically are made large enough to leave an imprint 11 to 18 centimeters (4.25 to 7 inches) deep, and the drum has six to eight teeth per row. A 5-centimeter (2-inch) gap is left between each row. The drum itself is larger than 50 centimeters (20 inches) in diameter. Any smaller, and the bottom of the imprint will become too rounded as the drum rolls over the ground. The drum must also have sufficient weight in order to make a proper imprint on the ground. If it is too light, the divot will be too shallow and may not provide the seeds with sufficient water to germinate and grow.

Imprinting can be done on flat or sloped land (within safe machine operating limits). Shallow soils or very rocky soils are not suitable for imprinters, however. Small brush will not interfere and will create a mulch for the site. If the brush is too dense, however, it will have to be cleared for the machine to work properly. The divots will last two years in sandy and clayey soils, and longer in loam. Treating up to four acres of land an hour, imprinters are a very efficient water-harvesting strategy in drylands.

Trees

Though not typically thought of as a water-harvesting feature, trees deserve inclusion in discussions on water harvesting. They should be planted in areas where water availability permits their growth. They will be beneficial in reducing wind and water erosion, stabilizing hillsides, building soil, improving water infiltration, contributing to nutrient cycling, improving crop and livestock health, increasing rainfall locally, reducing ground-level evaporation, humidifying air, and providing additional revenue streams for a site.

When rain falls on a forest, not all the water reaches the soil. A large portion of the rain is intercepted, first by the canopy of the forest and each tree's branches and trunk, and second by the leaf litter on the ground. In smaller rain events, when the canopy is not saturated with rainwater, the proportion of rainfall interception can be near 100 percent.

At first glance, it might appear that for water-harvesting purposes, trees should be removed from a site so that more water might then reach the soil and be infiltrated. When the broader hydrological cycle is considered, however, the value of trees is more easily understood.

In the event of water being intercepted by the canopy, the bulk of that water will evaporate and return to the atmosphere. In humid conditions, particularly in rainforests, it is not uncommon to see clouds of mist accumulating above the forest canopy. This is often misinterpreted as cloud formation as a result of leaf transpiration pumping water into the atmosphere. It is, in fact, intercepted water evaporating into the atmosphere. With the air re-humidified, this moisture can then travel downwind to fall as rainfall elsewhere. Increased local rainfall post-reafforestation is common.

Not only is rainfall intercepted by the canopy, so too is atmospheric moisture. Trees transpire moisture, cooling them. In the nighttime, the surface of leaves will cool faster than the surrounding air, leading to dew condensing on the leaves under the right atmospheric conditions. Excess dew will drip off the trees, leading to additional precipitation that is not measured by meteorological instruments. This captured dew also evaporates as described above, increasing humidity. Through interception of both rainfall and dew, trees can contribute to a leap-frogging of precipitation across the landscape. Intercepted water evaporates, traveling downwind to be re-intercepted as either precipitation or dew.

As a result of these processes, forest temperature and humidity changes are moderated. Daytime evaporation and leaf transpiration cools the canopy. This, in addition to the shade generated by the canopy, creates a cooler environment. In the nighttime, the movement of relatively warmer atmospheric moisture to the cooler surface of the leaf has a warming effect as condensation takes place. This increases the temperature in and under the canopy.

The interception–evaporation process also results in recordings of more humid air in forested regions. Increased humidity means less evaporative pressure at ground level, which, in turn, means less evaporative loss from the soil.

Canopy interception and detritus from the tree, including leaf litter and dead twigs and branches, means a reduced mechanical impact on the soil from the rainfall. Erosion is thus reduced under trees. The binding action of the roots contributes as well. Roots can help to stabilize an otherwise at-risk hillslope, reducing the risk of a slump or a slide.

The detritus from trees and the growth of tree roots help to loosen the soil, which increases water infiltration. The ecological niche created by trees also supports biological life in the soil in addition to being an attractor for wildlife. The nutrient contributions from the wildlife, in turn, help to feed the soil further, resulting in the formation of soil organic matter. In the case of nitrogen-fixing trees, the soil is further nourished, increasing soil fertility and fostering soil life. The soil life can lead to the formation of a crumb-like structure of aggregated soil particles. Clumped together with humus and plant and fungal exudates such as glomalin, these soils not only have a lower hydraulic conductivity, they also have a high water-holding capacity. Especially in temperate regions, the floor of undisturbed forests will have very soft, uncompacted soils that feel like a mattress when walked on.

Although a large portion of rainfall is intercepted by forest canopies, forests make both ground water and streamflow more consistent. More rainfall is intercepted, but rainfall as a whole increases, making for a more humid environment. Water that does reach the soil is more easily infiltrated, due to the formation of soil organic matter. In one experiment in New Zealand (Fahey and Jackson, 1997), a catchment of grassland was switched to pine, and the stream volumes from that catchment were measured. They found that the mean flood peaks were 55 to 65 percent lower from the catchment after it had been switched from grassland to pine forest. Increased interception helped to mitigate flood levels. It is reasonable to expect that increased infiltration, too, reduced runoff volumes that would lead to higher flood peaks.

Pattern planting

The tops of catchments should be planted, meaning the summits of hillslopes are best covered with trees. The trees will assist in the infiltration of water as described above, as well as in accumulating nutrients at the top of the site. The trees themselves can mine for minerals with their roots, making soil nutrients biologically available and concentrating them in the tissues of the tree. The trees will also attract wildlife, which carry with them a significant nutrient load that is left via their manure. Through the interception and reevaporation of rainwater described above, there is also an increased chance of greater total rainfall volumes occurring locally and regionally.

Trees at the summit also reduce erosion in the total system. Naturally, material from the interfluve moves downhill on a path toward the toeslope and beyond. Trees help to reduce this erosion and even contribute to soil creation at the top of the catchment. Slopes over 20° are forested for the sake of stability. A forested hillslope is much more resistant to sliding than a denuded one.

Foresting areas along waterways to provide a riparian buffer helps to protect streams and waterways from pollution and runoff. A 10-meter (approximately 30-foot) minimum buffer along rivers, streams, and lakes will keep out silt and pollutants as well as reduce evaporation and cool water temperatures.

This natural water- and nutrient-accumulation system has been put to use in systems such as the Japanese *satoyama* system of mountain agriculture. In this system, mountain summits and slopes were left forested and the footslopes terraced, reducing the risk of landslides. The trees also reduced the risk of flood to the lower portions of the watershed, which included cities. Leaves from the trees on the hillslope were collected annually in the fall and spread on the fields to fertilize them. Branches were cut and used for firewood or charcoal production. This system not only started generating biological production at the top of the catchment, it also was the starting point for the economy, with the products flowing downhill, as it were, from the mountainside satoyama systems.

Alley cropping and silvopasture

Contours can be used as guides for tree rows. This will assist in fighting erosion as well as adding windbreaks to a site. The trees themselves can be productive species, producing timber, fuel, or food. Nitrogen-fixing trees will add to fertility of the site. P.A. Yeomans advocated planting a band of trees along the keyline of sites to assist in water catchment as well as provide a windbreak for fields and pastures. Between rows of trees, crops can be planted. The trees are pruned and used as a green manure for the crops in some systems. The wider the spacing between trees, the longer that space can be farmed without competition from the trees impeding the crop yields. The greater the spacing, however, the more wind will be a factor for the crops. In temperate regions, available sunlight is also a major factor in deciding tree spacing. Looking at the average peak sun angle through the growing season versus the size of the tree at maturity will reveal how much shade you can expect. This can help you to determine the distance between rows of trees.

Temperate-climate cropping systems can benefit from the fertility offered by trees as well as the wind break that the trees offer. The tradeoff is loss of solar access in colder climates. Pollarding and coppicing the alley trees provides a work-around for this problem. Coppicing is the harvesting of certain trees at a height of 30 to 90 centimeters (1 to 3 feet) from the ground during the dormant season. The tree will grow back multiple shoots from the stump. The shoots are then harvested on a three- to ten-year cycle, depending on the rate of regrowth. This approach keeps the tree in a juvenile state and typically produces twice the biomass the tree would otherwise generate, while extending the life of the tree (provided the cut cycle is not too frequent). Pollarding is a similar approach, but the cutting takes place 180 centimeters (6 feet) or more from the ground (i.e., out of browsing range for livestock). As with coppicing, the cut cycle will depend on the rate of regrowth of the tree. It should be noted that not every tree will coppice, and not every tree that coppices will pollard, and vice versa. You will need to consult a database to find what trees are appropriate for your site. An alley three or more trees wide of alternating coppice and pollard trees will make a significant windbreak for crops between the alleys without significantly restricting the sunlight available for the crops.

The Burmese system of *taungya* (from *taung*, meaning hill, and *ya* meaning cultivation) is an alternative option for alley cropping. The taungya system grows crops in an alley-cropping regime between trees more closely planted together, whether for timber production or an orchard. The space between the trees is planted with crops until such time as the trees grow too tall, shading out the crop space. From there, the system is dedicated to tree production. During the cropping phase, care must be taken not to deplete the soil. Mulched and manured beds with nitrogen-fixing ground cover, such as white clover, should be employed. Rotation polyculture production should also be used to help prevent plant diseases that might later harm the tree production.

Pastureland, too, can benefit from trees. Animals seek shade on hot summer days. If adequate tree cover is available, animals can continue to graze while being shaded. In colder climates, trees also help to keep frost off the pasture. The trees also act as windbreaks, which helps to keep cold winds off livestock. This means a better feed–conversion ratio. Less food energy is burned to keep the animal warm, meaning that the livestock can put on weight faster.

As with alley cropping, the trees themselves can be for long-term timber production or can be fruit or nut producers. It is important to realize, however, that if you are planting nut and fruit trees on pasture for commercial purposes, you need to plan out harvesting. If the trees that come into harvest are scattered across a pasture or pastures, it will take extra time to harvest the crop. In the case of food-producing trees, you will need to clump together trees that fruit at the same time to make harvesting efficient.

In terms of crown coverage of the site, pastures can be up to 40 percent tree-covered without interfering with grass production. This planting can be done in rows on contour, with additional trees strategically placed as windbreaks and shade. Young trees still establishing themselves will need to be protected from grazing by the animals. If the tree species is palatable, animals will graze on any leaves and twigs they can reach and may also strip the bark off the trees.

Negarim microcatchments

In drylands with slopes under 3°, a diamond-patterned grid of small mounds designed to direct water toward trees is sometimes used. These system are known as *negarim* microcatchments, or net-and-pan systems. The top point of the diamond points uphill, and the bottom points downhill. The mounds on the lower portion of the diamond direct water downhill toward the lower point. Inside the lower point, there is a 40 centimeter (16 inch) deep depression where the water can collect. This is the point where a tree is planted.

While a V-shaped mound can be used for a single-tree planting, the diamond shape is used when planting multiple trees on more than one contour. The diamond pattern arises when one offset row of V-shaped mounds is placed above or below another row. In this way, a broad area can be established with trees, each with its own private water-harvesting system to support it. The mounds themselves can be formed in the same manner as swales, namely by cutting a trench on the uphill side and depositing the excavated earth on the downhill side. The trench will need to drain into the planting depression at the bottom point of the diamond. Dryland rain events tend to be both infrequent and large. The mound will have to be high enough to hold water in the microcatchment and must be planted with hardy grasses to protect against erosion. The following table gives figures for mound height above ground level for square (right-angled diamond) microcatchments against the slope of the hillside, assuming a mound with 5 centimeters of extra height above the uppermost corner of the diamond.

Table 7.1. Negarim dimensions

Basin size in meters	Height of mound for 1° slope (cm)	Height of mound for 2° slope (cm)	Height of mound for 3° slope (cm)
2 × 2	10	15	20
3 × 3	12	20	27
4 × 4	15	27	35
6 × 6	20	35	49
8 × 8	25	45	65
10 × 10	30	55	—

A variation on the diamond-shaped negarim system is an open V-shaped bund to capture water (see Figure 7.11). Excess water will spill over the ends of the V, reducing the volume of water that it can hold. If you are on rough, uneven land, or have minimal numbers of trees to plant, this approach can make a quick and easy solution. Adding riprap to the ends of the "V" will help to prevent erosion if excess water spills over the ends. This simplest approach is to dig a pit for the tree and place the excavated earth in a mound on the downhill side to increase the volume of the captured water.

FIGURE 7.11.
A typical negarim layout.

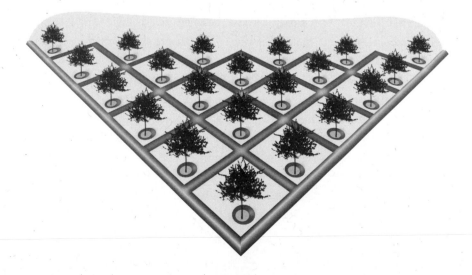

Spate irrigation

Spate (from the British term for a flash flood) irrigation systems harvest flood flows to support agriculture in drylands. Taking advantage of water flowing out of wadis, they capture large ephemeral pulses. These systems are used to irrigate cropland and rangeland, establish trees, recharge groundwater, and provide drinking water for humans and livestock. In situations in which high sediment loads in water flows would quickly silt up reservoirs, spate irrigation provides an alternative water-harvesting approach.

Spate irrigation does have a number of issues, however. Because of the sediment load carried by the water, the systems need regular, labor-intensive maintenance. The powerful nature of dryland flows also means that systems can become damaged or destroyed by large floods. Though the approaches used are simple, the hydrology of the overall system is complex and easy to get wrong. Additionally, because it is working directly with streamflow, it maximizes decoupling of the watershed. This requires a lot of effort in establishing community agreements regarding water rights, and social breakdown as a result of poorly planned systems that over-harvest is not uncommon. In regions where there is a long-standing tradition of spate irrigation, these water rights are more developed and accounted for—though not without the potential for problems regarding new projects or modifications to existing systems. In areas where spate irrigation is a recent import, large-scale projects have led to tense relationships along watersheds. For example, systems like Cubbie Station on the Culgoa River in the Darling River basin in Australia is licensed by the government to harvest 460,000 megaliters (372,930 acre-feet). This has reduced the flow of the Culgoa by one third and has halted downstream flooding, which both farmers and wildlife lower in the watershed rely on.

On a permaculture scale, spate irrigation projects would be well under the 1,000-hectare (2471-acre) systems that are considered small projects. Larger systems require careful, expert engineering, as well as community consultation. Even with experienced teams of engineers, success is not guaranteed. The challenge is to catch flows that are short and unpredictable in duration, and to design in such a way as to prevent larger, more destructive flows from damaging the system. Sediment buildup will occur and needs to be addressed.

Spate irrigation involves the diversion of part or all of a wadi's flood flow into a distribution system. The position in the landscape will determine what approach is used. Higher in the catchment near the wadi's exit, the land is steeper and the flow faster. In these situations, what is known as a spur-type diverter is used to direct a portion of the flow into the spate system (Figure 7.12). These are built a little like

FIGURE 7.12.
A spur-type diversion embankment and diversion channels. Spillways on channels will protect them against high-water levels.

highway off-ramps; they allow a portion of the wadi flow to enter the diversion channel while allowing the bulk of the water to continue to flow downstream. This diversion system is easier to build and can break apart in high-water conditions, helping to protect the rest of the system.

The other diversion system is the deflecting or diverting bund. These bunds are used lower in the catchment, where flows are shallower and more spread out. They are placed diagonally across the flow channel to take some of the force out of the flow of the water, and they are intended to divert all of the water into the spate system.

Water flows will cut into the wadi bed, lowering the bottom of the wadi channel. This will interfere with the diversion system and can eventually make it ineffective. You can improve the durability of the spate system by installing bed bars (also known as bed stabilizers) across the wadi channel. Bed bars consist of low weirs or gabions, perpendicular to the flow of the water. With the use of bed bars, care must be taken not to cut off the subsurface throughflow, thereby decoupling the entire catchment. Gabions won't carry this risk, but a concrete weir cemented to the bedrock will. On a smaller system, you will need to determine whether the added cost and labor is worth the improvement to the wadi bed. As mentioned for gabions and check dams in the "Dams" section (Chapter 6), bed bars must extend into the banks to prevent water from cutting around them.

Diversion bunds can be made more durable by adding a fuse plug along a portion of the bund. A fuse plug is a type of spillway that is made to intentionally break away in large flows for the purpose of protecting the overall system. In normal flow conditions, it will allow the passage of excess water. In large flow events, the fuse plug is swept away in the flow, allowing a greater volume to flow over the spillway.

With either diversion system, the water is carried along a diversion channel. Unlike every other type of water-harvesting earthwork, a high velocity of water flow is desirable. The reason here is that the water is so heavily laden with sediment that it will quickly choke the channel if the flow is slowed. Additionally, the minerals in the sediment are the main source of fertility for agriculture in these systems. Ideally, you will want the sediment to continue to flow to the fields, where it will assist in plant growth.

Installing a spillway in the diversion channel will help to protect it in the event of high flow without adding too much in terms of additional cost. The spillway is located on the wadi channel side so that excess water is returned to the main flow channel.

The diversion channel directs the water into fields contained by bunds to hold the water. Typically, these are entirely enclosed with pipes connecting fields, but they can be open systems, such as variations of the trapezoidal bunds described in the "Swales" section (Chapter 7). Runoff from open rock surfaces can also be channeled to such bunds via concrete gutters.

On a small scale, waterproof liners or clay barriers can be used to create subterranean water storage to hold a portion of the water that seeps into the ground for agriculture. This would be over only a small area to avoid decoupling too much land and interfering with groundwater recharge.

Although the systems above describe fast-flowing diversion channels to carry sediment to fields, orchards, rangeland, and groundwater recharge reservoirs, you can build a system that moves water to a dam or cistern, provided the sediment load in the water is not too great. If there is a section that is deep enough and slow enough to allow large particles to drop out, a diversion channel can be built to direct a portion of water to a cistern or a dam. In either case, shading the storage will help to prevent evaporation in drylands.

Similar to dam construction when there is sediment flow, a silt trap will need to be constructed. This can be an earthen pit, though a concrete settling tank with baffles between two or more settling basins is more effective and easier to maintain. With such a system, a bed bar would be placed downstream of the diversion channel. If the water approaching the deeper zone can be slowed with weirs to encourage more sediment to drop out, this will help to reduce silt buildup in both the zone around the diversion channel, and the silt trap. Cleaning of the silt trap will be necessary after each spate flow.

River training

River training can be used on its own or in conjunction with spate irrigation systems. River training is a complex subject. For our purposes, we will look at a few simple approaches to guiding river flow.

The behavior of a river will depend in large part on the slope of the river bed. Any work done will be on rivers with a bed slope of less than 4 percent, or 2.29°. There are ultimately two basic options with river training: spreading water or concentrating water. Spreading slows down the flow and will allow any sediment to drop out of the water. Concentrating will speed up the flow and cause greater erosion.

Spreader banks can be effective in drylands for both distributing water and reducing the erosive power of water. This may come in the form of a low weir, a gabion, or even a line of rocks used to divert water away from its downstream path, and onto a path that is closer to contour. Keeping the fall of the spreader to within 1:500 slope will help to reduce erosion and help to prevent undercutting of the spreader.

The way to create a small-scale erosion-control spreader-bank approach suitable for rills or channels up to 30 centimeters (1 foot) deep is to make a fan-shaped band of rocks that spread the water in a radial arc downstream. The arc shape creates a catchment bowl upstream so that no water entering the top can spill around the sides. Erosion is halted, and the water is spread out radially when exiting. Such systems are helpful in checking erosion in the transition from footslope to toeslope.

The inverse of this structure can be used at the top of the catchment in valleys to control erosion at the highest points. Where rills or small channels begin to form

FIGURE 7.13.
Concave and convex "media luna" used to control erosion.

in a head slope, an arcing rock layer is laid out on contour to cover the desired rills and channels; then additional lines of rock are placed on contour above the last line. Both these spreading and collecting systems, termed *media luna*, were conceived by Van Clothier and further developed by Craig Sponholtz (see Figure 7.13).

A spur dyke can be placed within a stream bed from one bank, extending partially into the river. This will slow water along the bank. This can be angled perpendicular to the flow, angled with the end pointing upstream, or angled with the end pointing downstream. Within a series of two to five spurs, a groin is formed. Groins are typically used to protect banks from erosion while increasing the water velocity, and thus depth, in the main channel. The space between the spurs will create a groin field in which silt will deposit. The spurs themselves can be impermeable, or they can be permeable and submerged or extending above the waterline. It should be noted, however, that the below-water depth for submerged spurs is a more important factor than total height of the spur. Silt deposition decreases as the depth of the spur increases.

Spurs placed in line on either side of a river will function as a wing dam, concentrating the flow in the center, leading to faster, deeper water. Whether creating wing dams or a groin, the spurs must extend into the bank so that no water cuts into the bank and behind the spur. This is especially crucial if the spur tip is angled upstream.

On a small scale, you can place a single layer of rocks across a straight section of river bed, much like a bed bar. This creates a shallow section known as a riffle, which takes energy out of the river flow. Similar to bed bars or gabions that cross wadis, these rock layers must extend up the banks, otherwise water will cut around the edges. Pools will then naturally form on the downstream side of the artificial riffle. These artificial riffles were originally developed by Bill Zeedyk, which he described as "one-rock dams."

If the soil along a stream is eroding into a cut, forming dropoff that creates a small waterfall, the erosion will continue retrogressively upstream if left alone. In these cases, cut the edge of the dropoff back at a slope of 1:2. A splash apron will be necessary. For this, dig a cutoff trench right across the stream bed and into the bank at a distance downstream three times the height of the drop downstream from the drop. Fill the cutoff trench with rock to a height of 5 centimeters (2 inches) above the stream bed. Place large rocks that are approximately half as high as the dropoff just upstream from this rock layer in the cutoff trench so that water flowing over these rocks will be met by the rocks in the cutoff trench. Line the entire inside with rocks, including up the banks and up the 1:2 slope where the dropoff was, being sure to offset the rocks as you would brickwork to avoid forming a channel that could erode. Fill the space as much as possible with rocks, avoiding bare patches. A one-rock dam is then placed across the stream bed at a distance six times the height of the dropoff from the position of the dropoff. This erosion-control approach was developed by Bill Zeedyk and the community at the Zuni Pueblo, and has been called the "Zuni bowl."

Diversion drains

There can be situations in which waterlogging is a problem you want to avoid. For instance, there may be an existing building in the toeslope that is prone to flooding during wet periods. There may also be cases where you want to prevent saline seep by intercepting salty throughflow. In such instances, water can be diverted and drained away to help prevent waterlogging and/or saline seeps.

On the hillslope, you will want to cut in diversion ditches that fall off to one side or the other of the wet area downhill. The trench should be 500:1 or less, depending on how erosive the soils are. The aim is to remove water, not soil. Unlike swales, the trench will be compacted to restrict infiltration as much as possible. The depth of the trench will also be deeper than that of a swale—typically from 1.5 to 2.5 meters (4.9 to 8.2 feet) deep. You will ideally want to cut down to intercept the subsurface throughflow, not just intercept runoff.

References

Anschuetz, J., A. Kome, M. Nederlof, R. de Neef, and T. van de Ven. *Water Harvesting and Soil Moisture Retention*. Wageningen: Agromisa Foundation, 2003.

Callow, J.N, and Smetten, K.R.J. "The effect of farm dams and constructed banks on hydrologic connectivity and runoff estimation in agricultural landscapes." *Environmental Modelling & Software*. Vol. 24, Issue 8. (2009) doi: 10.1016/j.envsoft.2009.02.003.

Chang, Mingteh. *Forest Hydrology: An Introduction to Water and Forests*. Third Edition. Boca Raton: CRC Press, 2013.

Critchely, Will, Klaus Siegert, and C. Chapman. "Water Harvesting Techniques." *Water Harvesting: A Manual for the Design and Construction of Water Harvesting Schemes for Plant Production*, FAO Corporate Document Repository. fao.org/docrep/u3160e /u3160e07.htm.

Crozier, Carl. *Soil conservation techniques for hillside farms*. Washington DC Peace Corps, 1986.

Doer, B.D. *Land imprinters, Section 8.2.7., US Army Corps of Engineers Wildlife Resources Management Manual*, Technical Report EL-86-43. Final Report, Department of the Army, US Army Corps of Engineers: Washington, DC, 1986.

Fahey, Barry, and Rick Jackson. "Hydrological impacts of converting native forests and grasslands to pine plantations, South Island, New Zealand." *Agricultural and Forest Meteorology*, 84 (1997).

FAO Corporate Document Repository. "V. Classification of Terraces and Ditches and Selection Criteria." *Watershed Management Field Manual: Slope Treatment Measures and Practices.* fao.org/docrep/006/ad083e/AD083e06.htm.

FAO Corporate Document Repository. "VI. Continuous Types of terraces (Bench Terraces)." fao.org/docrep/006/ad083e/AD083e07.htm.

FAO Corporate Document Repository. "VIII. Transitional Types of Terraces (Convertible Terraces and Intermittent Terraces)." *Watershed Management Field Manual: Slope Treatment Measures and Practices.* fao.org/docrep/006/ad083e/AD083e09.htm.

FAO Corporate Document Repository. "IX. Terraces for Gentle Slopes of Rangeland (Broadbase Terraces and Natural Terraces)." *Watershed Management Field Manual: Slope Treatment Measures and Practices.* fao.org/docrep/006/AD083E/AD083e10.htm.

Law, Ben. *The Woodland Way: A Permaculture Approach to Sustainable Woodland Management.* East Meon: Permanent Publications, 2001.

Mollison, Bill, *Permaculture: A Designers' Manual.* Tyalgum: Tagari Publications, 1988.

Nair, P.K.R. *An Introduction to Agroforestry.* Dordrecht: Kluwer Academic Publishers, 1993.

Owes, Theib Y., Dieter Prinz, and Ahmed Y. Hachum. *Water Harvesting for Agriculture in the Dry Areas.* London: CRC Press, 2012.

Plant & Soil Sciences eLibrary. "Chapter 3 Soil Water." *Irrigation Management.* croptechnology.unl.edu/pages/informationmodule.php?idinformationmodule =1130447123&topicorder=3&maxto=13&minto=1.

Rankka, Karin, Yvonne Andersson-Sköld, Carina Hultén, Rolf Larsson, Virginie Leroux, and Torleif Dahlin. *Quick clay in Sweden.* Report 65, Linköping: Swedish Geotechnical Institute, 2004.

SearNet. "Negarim Micro-Catchments." International Rainwater Harvesting Alliance, 2011-08-12. irha-h2o.org/?p=651.

Sheng, Ted C. "Bench Terrace Design Made Simple." 12th ISCO Conference, Beijing 2002. tucson.ars.ag.gov/isco/isco12/VolumeIV/BenchTerraceDesignMadeSimple.pdf.

Shrestha, A.B., G.C. Ezee, R.P. Adhikary, and S.K. Rai. *Resource Manual on Flash Flood Risk Management : Module 3—Structural Measures.* Kathmandu: International Center for Integrated Mountain Development, 2012.

Spate Irrigation Network Foundation. spate-irrigation.org/.

Spate Irrigation Network Pakistan. "Command Area Improvement and Soil Moisture Conservation in Spate Irrigation. Practical Notes #4." ocw.unesco-ihe.org/.

Spirko, Jack. "Don't Try Building Hugel Swales—This is a Very, and I mean Very Bad Idea." 2015-11-06. permaculturenews.org/2015/11/06/dont-try-building-hugel-swales -this-is-a-very-and-i-mean-very-bad-idea/.

Sponholtz, Craig, and Avery C. Anderson. "Erosion Control Field Guide." 2010. watershedartisans.com/Erosion_Control_Field_Guide.pdf.

Takeuchi, K., R.D. Brown, I. Washitani, A. Tsunekawa, and M Yokohama. *Satoyama: The Traditional Rural Landscape of Japan*. Tokyo: Springer, 2003.

The Imprinting Foundation. imprinting.org/.

van Steenbergen, Frank, Philip Lawrence, Abraham M. Haile, Maher Salman, and Jean-Marc Faurès. *Guidelines on Spate Irritation*. FAO Irrigation and Drainage Paper 65. Rome: FAO, 2010.

van Steenbergen, Frank, Abraham M Haile, Taye Alemehayu, Tena Alamirew, and Yohannes Geleta. "Status and Potential of Spate Irrigation in Ethiopia." *Water Resource Management*, Vol. 25 (2011) doi: 10.1007/s11269-011-9780-7.

Waelti, Corinne. "Micro Basins." sswm.info/print/1519?tid=519.

Yeomans, P.A. *The Challenge of Landscape: The Development and Practice of Keyline*. Sydney: Keyline Publishing, 1958.

Yeomans, P.A. *Water for Every Farm*. Second Edition. Sydney: The K.G. Murray Publishing, 1968.

Zeedyk, Bill, and Van Clothier. *Let the Water Do the Work: Induced Meandering, an Evolving Method for Restoring Incised Channels*. White River Junction: Chelsea Green Publishing, 2009.

Applying Permaculture Strategies

Chapters 6 and 7 described a set of techniques that can be applied for the purpose of harvesting water. But just a as a set of woodworking tools won't guarantee the creation of a sturdy, functional piece of furniture, so too a set of techniques won't guarantee a greener landscape. The trick is to create a system that does not underperform and is not dangerous.

Goal setting, planning, adjusting

As highlighted in Chapter 4, the first step in any project is to clarify what it is you are trying to do. If the goal of the project is not clear, then a desirable outcome will be much harder to reach. Failure to adequately articulate the goal of a project is the biggest error that novices and professionals alike make. Be sure you can easily answer what it is you are trying to do and why. If you cannot explain your motivations or your intended actions, then you are starting out on shaky ground.

After you determine your goal, you are ready to make a site plan. The approaches and ideas presented below will assist you in this. It is important to keep in mind, however, that you can never have perfect information about your site or what will happen to it in the future. As a result, your plan will be flawed. The plan is a best guess for how to proceed, but it will be imperfect.

Because of this, issues will arise in the process of both planning and implementation that will require a modification to your plan (and possibly your goal) in light of feedback and information you gain as you proceed. If, for instance, you discover unstable conditions on a site where you had planned to build a dam, you will have to adapt to this information and alter your plan.

The techniques presented in Chapters 6 and 7 describe the purposes of various types of earthworks and how they are constructed. Depending on your confidence level and experience, however, you may not be comfortable executing them yourself. This is not a problem. As a designer, you can design the big picture and turn to experts for help. There are cases in which you should call on the consultation of engineers and hydrologists.

Zone planning

Zone planning follows a simple directive: Plan action around energy, not energy around action. In other words, don't build a system without considering the amount of energy required to use and maintain that system. We'll look at incorporating crop-production systems with open water storages below, but consider how ineffective such a system would be if it were located in a pond or dam on a site that took ten minutes to reach from the main building or functional center of the site. In such a case, you would either need to resign yourself to spending twenty minutes or more a day in traveling to the site or use a horse or all-terrain vehicle to reach the site. In either event, it would require an energy expenditure that could have been avoided with better planning. Placing elements out of their optimal zones creates the need for either greater labor or external energy inputs. The physical landscape is the prime contributor when it comes to deciding where to place earthworks, but they also need to be placed so that they carry out functions on a site efficiently.

Permaculture divides a given site into six zones. The first, zone 0, is the house or main building itself, including any attached structure such as a trellis. In this case, a cistern incorporated into or attached to a house would be in zone 0. Zone 1 contains elements and systems on the site that require near constant attention. Zone 1 is visited daily, typically multiple times a day. Examples would be home gardens, seedlings, and laying hens. Zone 2 includes site systems and elements that require visits from once a day to once every two or three days. Typical zone 2 elements include market gardens, spot-mulched orchards, and animal shelters. Zone 3 is typically visited from once a day to once a week. Large-scale water storages, grazing animals, broad-scale farming, silage storage, barns, and agroforestry systems are the sort of elements found here. Zone 4 contains elements such as rangelands, timber production, and water-harvesting features. Zone 5 is essentially a wild zone. It is occasionally foraged, and the woods there might have some minimal maintenance, but the zone is mostly left alone (see Figure 8.1).

Zones are determined based on the amount of energy needed to reach them. On relatively flat ground, this might be based on distance from the most frequently used door. The greater the variety in the site's topography, the more the slope of

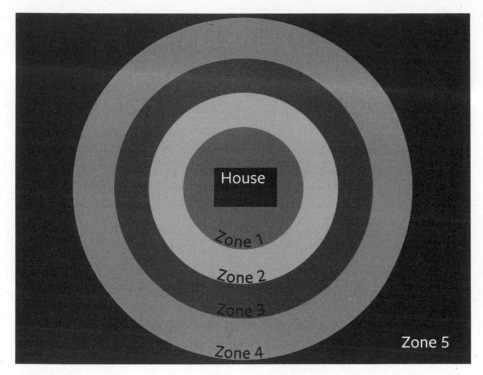

FIGURE 8.1.
Zone planning based on energy requirements to reach each zone. In the real world, the boundaries of the zones will be irregular and will be determined by the site's geography as well as elements on the site that affect movement.

the land will determine the relative boundaries of the zones. Placing zone 1 or 2 elements in relatively difficult places to reach creates more work than is necessary.

With respect to the water-harvesting systems you apply on a site, zone planning will help you to place them in a manner that allows them to carry out their assigned functions without reducing the sustainability of the site by requiring additional energy inputs due to the placement of the systems. Considering the example given above, you would want to avoid designing a garden that would require a 20-minute round-trip to visit that garden. While the functioning of the garden might be enhanced by connecting it with the open water storage, the gains would be lost, given the difficulty in reaching the garden.

Sector planning

Every site will be subject to incoming energies and inputs that have a direction behind them. The course of the sun follows a sun path that varies with the Earth's axial precession. In temperate regions, this is a major determining factor in the placement of elements on a site. The prevailing winds have directionality. There

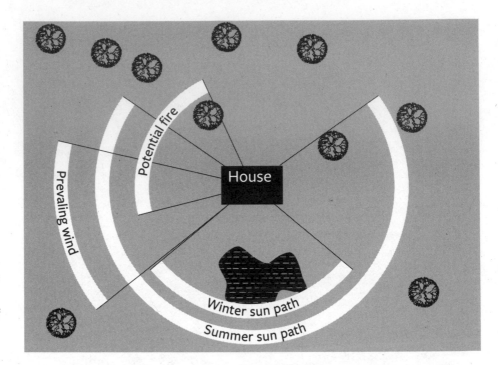

FIGURE 8.2.
Sector analysis of a site. Sector analysis helps to illustrate the directionality of relevant energies affecting a site.

will be a general area from which the winds come. Very often storms and other distinct weather systems will have a separate direction from the prevailing wind. Wind direction, combined with slope and existing vegetation, can help to predict from which direction wildfires might spread to the site. If floodwater collects on a site from water flowing onto the property from a certain direction, this can be considered the direction from which floods come. Any external force or input that enters the site can be assigned a sector if it has a reliable direction. From this, you can make an overlay onto a site map to help visualize the directional impacts on the site (see Figure 8.2).

A concrete example for this could be the placement of open water storage so that it either maximizes or minimizes solar gain on that storage. Maximizing solar gain can be helpful when you want the water storage to mitigate surrounding temperatures by creating a microclimate that extends the growing season. If you want to minimize evaporative losses and algae growth, minimizing solar gain would help. If you are in a region where wildfire is a threat, open water storages can be placed as a barrier between buildings and the most likely path of an oncoming wildfire. The water storages can also serve as a firefighting tool—either through flood flow irrigation of the site or through the use of pumps used to fight the fire.

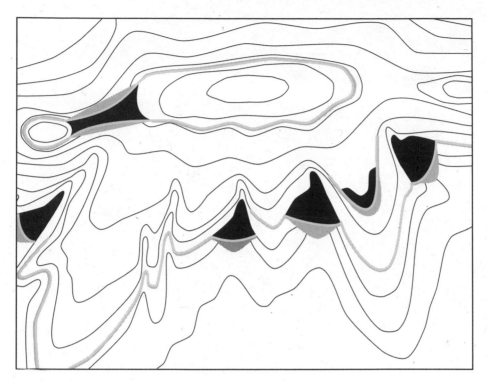

FIGURE 8.3.
Swales can increase the catchment for a dam. They can also integrate spillways and connect multiple dams.

Stacking functions and functional connectivity

Recall from Chapter 4 that stacking functions means making an element on a site carry out more than one task, and functional connectivity means linking two or more elements in a system together in such a way that they can then carry out some function together.

In Chapter 6, you saw how swales can be used to increase the catchment areas for a dam. The swale stretches laterally across the landscape, capturing water that would otherwise run off and away from a dam (Figure 8.3). The swale is open at the dam end, allowing collected water to flow into the dam and fill it. The same swales that fill the dam can have built-in spillways that allow excess water to escape the dam in a desired location. And because the swales will backfill as the dam's water level rises, they will allow more water to infiltrate along the swales, and the total capacity of the dam system will be increased. It is not uncommon to add hundreds of thousands of liters of capacity, or more, to a dam in this way.

There is a lot of versatility with swales. By making the bottom of a swale wider, it can double as a pathway or tractor path during dry conditions, though this will compact the bottom somewhat, reducing the swale's capacity to infiltrate water.

FIGURE 8.4.
Swales can also assist in the capture and distribution of nutrients. In this example, nutrients from cattle manure are spread more evenly across the landscape as water flows into the swales.

The very nature of swales is to slow the flow of resources—namely water—downhill and off a site. This flow can include the flow of nutrients as well (Figure 8.4). This might be done broadly by having an animal system above a swale. The nutrients from the animals gradually flow downhill and into the swales, where they are more evenly distributed across the landscape as water enters and fills the swale.

A spinoff of this approach is the placement of a chicken coop or rabbit hutch directly over a swale. Rabbit hutches typically have a wire cage floor to allow the droppings to fall through. A similar design in a chicken coop would also allow the droppings to fall through and into the swale. As the swale fills with water in a rainstorm, the nutrient load can flow along the swale to some extent. In such a case, excess droppings would build up and would need to be cleared out and composted, though this is a necessary step with a more typical coop or hutch in any event.

Islands in ponds and dams can make a convenient location for chicken coops. Chickens will not cross water, so containing them on an island is a simple matter. The island also affords additional protection for the chickens. While predators will cross water to get at chickens, the water does provide an additional barrier that makes it more difficult to access the birds. A potential drawback with this approach is that the nutrients from the chickens can spur algae growth in the water.

A pond or dam will create a microclimate in one of two ways. Water holds heat much longer than air can. While springtime temperatures are pushed lower by the presence of a body of water, the water in a pond or dam, which is relatively shallow, will heat enough in the spring that it reduces the risk of nighttime frosts. The water will also cool more slowly in the autumn, increasing the temperature of the surrounding area and similarly reduce the risk of frost. This can extend the growing season for land adjacent to the pond or dam. The effect of a cooler body of water on the surrounding air reduces temperatures in the vicinity of the water on hot days. In this way, both hot and cold temperatures are moderated by the water.

Open water storage also reflects light. A pond placed near a home, for instance, will increase the solar gain in the house, increasing inside temperatures. The same effect will be experienced with trees where there is a reservoir between the sun and the trees. Here, the body of water can carry out the added task of increasing the available light.

Variations in structure and shape

The Mesoamerican system of the chinampas (developed independently in other parts of the world) is a highly productive approach to growing crops. In the Valley of Mexico, areas were staked out in the shallow waters of Lake Texcoco, Lake Xochimilco, and Lake Chalco. Chinampas consisted of a gridwork of sections of lakebed that were staked off, then filled with cut vegetation and soil from the lake bottom to create beds for agriculture. The result was blocks of fields with water between them, which provided access and irrigation. The water also served to moderate temperatures within the fields. With up to seven harvests per year, these were among the most productive systems in history.

This pattern of alternating blocks of land and water is often adopted along the edge of permaculture systems in a modified form. The edges of ponds or dams (excluding the dam wall) can be cut back in strips to leave a crenelated edge of strips of land that extend into the water (Figure 8.5). These strips can then be used for agriculture or agroforestry. The proximity to the water will help to moderate temperatures, elongate the growing season, and ensure adequate irrigation. The addition of trees on chinampas can also help to reduce evaporative losses. Soil that is dredged from the adjacent bays of water provides fertility for the crop beds or trees. The bays between the strips create varied protective habitat for the aquatic life in the pond or dam.

As mentioned in Chapter 4, evaporative losses from a pond or dam can reach 10 to 20 percent of the total volume in humid regions and 70 percent in drylands, with greater losses a possibility in semi-arid and arid regions. Being shallow bodies of water, ponds and dams heat up relatively quickly, which, in turn, increases the volume of water lost to evaporation. The greater the surface, the more evaporation

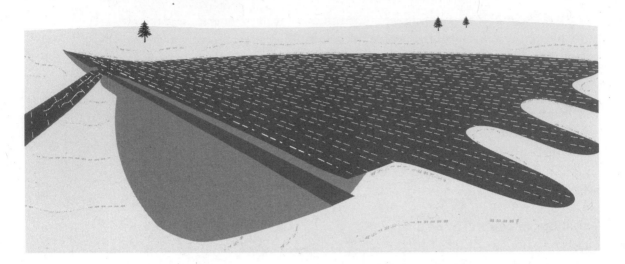

FIGURE 8.5.
The edges of ponds or dams can be crenelated to mimic the form of chinampas. These extensions of the reservoir's edge create microclimates, make irrigation easier, and take advantage of the nutrient-rich soil formed in shallow water.

takes place. One strategy to combat evaporation is to make smaller, deeper ponds and dams in drylands. There is a limit to the possible depth, however, as the angle of repose for soils is typically lower underwater when the soil is saturated than when the earth is below the saturation point. In other words, if you cut the edges of the reservoir too steep, they will just erode and fall inward and reduce the depth and volume anyway.

Trees planted around the perimeter of a reservoir (though never on a dam wall) can help to provide shade to reduce evaporation. This is not the only approach available, however. A network of wire trellises can be built to shade all, or a portion, of a reservoir. Wires can be staked on the shoreline of a pond or dam, and running vines allowed to cross the water, as shown in Figure 8.6. Vines may not grow long enough to reach and cover the middle of the reservoir. Should you wish to increase the coverage in these situations, you can tether a floating planter to the wire. Train the vine up the tether so that it can reach the trellis wire. The tether will have to have enough slack to allow for drops in water level. It will also need a wick in the bottom that allows water into the planter for the plants.

An approach similar to this has been studied by researchers at the University of Costa Rica and the Ensenada Center for Scientific Research and Higher Education. The method they studied was to turn lake surfaces into productive spaces by growing container plants on floating trellises. Their research found that the evapotranspiration from plants is only slightly greater than evaporation. There are cases, however, in which plants can decrease wind and spray, which leads to a decrease in evaporation. Furthermore, the floating platforms used to grow plants

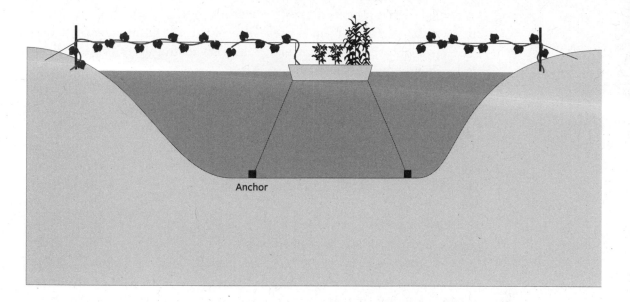

Anchor

can contribute additional shade, reducing evaporation. With regard to losses from evapotranspiration, if losses enhance food production, they are acceptable losses, particularly if the water would otherwise just have evaporated.

The lake production studies showed poor results when attempting to grow terrestrial crops with their roots directly in the water. This was due to low nutrient levels. When plants are placed in a soil medium with a wick to draw moisture from the body of water, however, they were able to thrive. The research team discovered an additional advantage of these systems in that they are harder for pests to reach. Pest and weed problems are very uncommon, unless they are inadvertently transported from land to the aquatic beds. In comparative studies conducted on whitefly and leaf miner numbers in bean, tomato, and bell pepper plants, researchers found that the aquatically grown plants had no whiteflies, beans had 96 percent fewer leaf miners, and tomatoes had 93 percent fewer. (Neither the terrestrial nor aquatically grown peppers suffered from leaf miners.)

Crop yields were also significantly higher in plants grown on the water compared to plants grown on land. The research was conducted at La Virgen Lake and Lake Managua in Nicaragua and at Lake Arenal in Costa Rica. The average lake yield compared to the land yield was 252 percent higher for beans, 195 percent higher for lettuce, and a phenomenal 1196 percent higher for tomatoes. The experiment at La Virgen Lake also studied rice, which had a yield 143 percent higher on water than the terrestrially grown crop.

FIGURE 8.6.
Vines trained over water can help to reduce algae and evaporation. Crops can also be grown from floating platforms anchored in a pond or dam.

Similar floating planters can be used to float aquatic grasses in the water, which can be anchored to hold them in a relative position. The plants will help to draw nutrients out of the water, which, in turn, helps to control algae. Algal blooms are the result of excess nutrients and access to sun. Removing one or both of these factors will reduce the amount of algae in the water. Duckweed will do both of these jobs in addition to reducing evaporation. In temperate climates, algae gets a head start in the spring and will establish faster than duckweed. If you want to control algae with duckweed in temperate regions, keep an aquarium full of duckweed over the winter, then add it to the water in the spring three to five weeks after ice out. This will give the duckweed a chance to establish faster than algae, which could otherwise take over.

If algal blooms have already established, barley straw can be used to inhibit algal growth. As barley straw aerobically decomposes, it releases a chemical that restricts the growth of algae, eventually bringing algal growth under control. Aerobic decomposition is key in using barley straw. Place the straw in a burlap bag with floats to ensure that the bag stays on the surface, where it will have exposure to the air. The amount of surface area in the reservoir determines the amount of straw needed. You will need 50 grams of straw per square meter of surface area (6 square feet per ounce of straw). Alternatively, there are now commercially made pellets that contain a mix of barley straw and dormant bacteria. The barley inhibits the algal growth, and the bacteria consumes nutrients in the water, choking out the algae. These pellets are safe for frogs, fish, and aquatic insects.

Flow

Open water storages can be made more conductive to biological diversity by adding variation to the depth of the reservoir. Building a shallow bench is a straightforward matter of leaving a section unexcavated in the implementation stage. Shallow areas receive more light, are quicker to heat up, more easily support the growth of aquatic plants, and provide a refuge for frogs, smaller fish, and aquatic insects. The increase in biological activity, in turn, attracts birds and other wildlife to the shallow regions of the reservoir. This intense biological activity accumulates nutrients and leads to the rapid formation of soil at the bottom of the reservoir. This can then be dredged out to provide fertility for terrestrial systems. Attracting wildlife to a site will also mean an increase in imported fertility. Though often considered pests with only the potential to lead to crop losses on a site, wildlife carry with them nutrients from off site. Natural systems are flow systems, meaning that they survive by the movement of resources through the open borders of the system. To cut off a site from this

Tarps are staked over the ditch to force water to back up and flood.

FIGURE 8.7.
Flood-flow irrigation.

flow of resources is to limit the availability of nutrients necessary for all biological activity on the site, including the systems that you want to encourage. Open water storages are a focal point for the movement of wildlife through a system and should be considered assets that contribute to site fertility.

In P.A. Yeoman's keyline system, he would often connect the landscape with a series of dams at the keypoints of primary valleys or head slopes, linked by a diversion ditch with a sluice gate to have control over the flow. Typically, the keypoint on the head slope in each primary valley is higher than the previous one as you move up the main ridge in the system. This allows for a gentle gravity flow for the system from the most highly elevated valley to the lowest one.

On flatter land in the toeslope, where erosion will not occur with water flowing over the surface, Yeomans would have nearly level channels that could be filled with water that would spill over the bank and flood an enclosed section of field or pasture (Figure 8.7). The flooded sections are called bays, and each bay is enclosed with a bottom bund placed on contour and side bunds that run perpendicular to contour. Waterproof tarps are placed across the channel, with the upstream edge staked to the bottom of the channel at either leading edge of the tarp. A cord is attached to the stake and the corner so that the stake can be pulled out and the corner lifted to allow water to pass under the tarp in a controlled manner. The downstream end of the tarp is attached to a pole that rests across the top of the banks of the channel. As water is let into the channel, the flag causes the water to back up until it spills over the downhill bank and into the irrigation bay. For each bay along the run of the channel, a flag will be placed across the channel to back up the water and flood each bay in turn. This allows for rapid and thorough irrigation of the land.

Efficiency

As mentioned in Chapter 4, dams for use in irrigating crops have a greater impact on the watershed because more of the water that flows onto the site stays on the site. Crop-irrigation dams are typically drawn down more than dams for livestock, which in turn means that there tends to be less overflow flowing over spillways and continuing downstream in the watershed compared to livestock dams. One way to have less impact on the watershed is to make use of efficient irrigation. Sprinklers are versatile, convenient, and wasteful. In conditions of high temperatures and low humidity, evaporation can reach as high as one half of the volume of water used. Additionally, excluding low-energy, precise-application systems, sprinklers lose a lot of water to both interception on leaf surfaces and irrigation of space between plants.

In contrast, drip irrigation, while more expensive and labor intensive to set up, greatly reduces evaporative losses by delivering water directly to the plants' root zone. Drip irrigation improves overall water use efficiency and increases yields. The addition of mulch between rows and around plants, as well as establishing ground cover, also helps to reduce evaporation. If you must use a sprinkler, time your waterings to reduce evaporation, in early evening or early morning when temperatures are lower and relative humidity is higher are better times to water.

Where to start

You can gain efficiency from your overall system by starting where your action will have the greatest impact for the smallest effort. This means starting as high in the watershed as you can. At the top of the watershed, small efforts to capture and infiltrate runoff will help everything downhill, starting with the first rainfall. The easiest way to go about capturing water at the top of a catchment is with trees. Hilltops covered with trees will reduce evaporation, improve the hydraulic conductivity and water-retaining capacity of soils, capture occult precipitation (condensation and interception of fog), help to increase local rainfall through evaporation of intercepted rainfall, and help to build soil fertility that will have downhill benefits as the nutrients are slowly carried downhill.

You may also have the opportunity for swales, rock walls, and gabions to assist in erosion control and water infiltration. Ripping will greatly help, given appropriate soils, as outlined in Chapter 7. If there is enough of a catchment to support a pond or dam near the top of a catchment, it will become a magnet for wildlife, and the nutrients they carry in will accumulate and gradually flow downhill.

Most sites you work on will not be at the top of a catchment. With these sites, it is still best to work from the top down in most cases. The general pattern is to use smaller efforts at the higher parts of the site and larger earthworks as you move downhill. Irrespective of where you are on a site, trees are a good idea unless they will actively interfere with other systems on the site. And even on a site where earthworks would be unsafe to attempt, trees will help with water harvesting while stabilizing the earth against sliding and slumping. Remember that alley-cropping agroforestry systems and silvopastures typically increase productivity when the ratio of trees to field, or trees to pasture space is correct. (See Chapter 7.)

Infiltrating water higher in the landscape means that the water will be forced to travel a slower path off site. In terms of broader watershed health, this generally means a lower total flow volume, but a more consistent flow of streams and rivers via groundwater recharge. Allowing maximum runoff will increase river flows but will add to erosion and greater downstream silting issues, and will decrease the amount of time water is available before it hits the ocean. In many cases, even significant rivers can become ephemeral, having massive flows during large rain events, then ceasing to flow during dry periods. The only instances in which increased infiltration is not desirable are cases in which downhill waterlogging could create a problem, saline seep could occur (see the "Swales" section in Chapter 7), or there is a risk of landslide—a problem we will examine in the next chapter.

Desert communities tend to develop around mountains, inselbergs, and other areas with varying elevation. The elevated areas serve as a macrocatchment that feeds agriculture beneath it. Even in areas of more plentiful rainfall, the elevated areas intercept rainfall, which communities in the footslope and toeslope take advantage of. The overall pattern of approaches such as Yeoman's keyline system is to intercept and store a portion of that water at the highest point possible for use downslope. The patterns of use that have emerged around the world have done so to adapt to the flow of water across the land.

Case Study: Circle Organic ridge point dam

Circle Organic is a newly established organic farm near Peterborough, Ontario, Canada. With drought being a regular feature most summers, and new record droughts occurring twice in the past decade, the farm owners approached me to design and implement a farm dam for them.

Two locations seemed ideal—one for a contour dam, the other for a keypoint dam. Having young children, the owners thought that neither location was

FIGURE 8.8.
Clay being loaded
into the keyway,
15 cm (6 in) at a time,
then compacted.

acceptable, as they would not be able to see the water from the house and thus couldn't keep an eye on their children's safety. This left only one option. The site has a ridge adjacent to the house and in front of the barn and greenhouse. This meant a ridge point dam would be built.

The dam itself is close to the house and rests inside zone 2. It is near the top of the site, meaning catchment is at a minimum. To assist in filling the dam, the plan was to install 760 meters (2,500 feet) of swales.

Unfortunately, ridge point dams are the most expensive to build as they have a maximized wall length for the volume of water they hold. In the case of Circle Organic, the keyway was 52 meters (170 feet) in total. A keyway that long requires a lot of clay. As in all of Ontario, the site has experienced recent glaciation. This means that there are many layers of deposits on site—clay deposits but also sand, silt, and gravel. This meant that mining for clay was somewhat challenging. There was also a section of the keyway that contained rather silty material that was somewhat suspect. Digging deeper did not reveal better material.

Though the wall of the dam was not very high, the dam was dug to a depth of 3.5 meters (12 feet), both to meet the desired capacity for the dam and to find adequate clay for the keyway. The key was filled with the best available clay in

FIGURE 8.9.
Good clay being
spread across the
inside of the dam
wall.

15-centimeter (6-inch) lifts and compacted several times with a padded-drum compactor before adding additional lifts. A 30-centimeter (1-foot) layer of good-quality
clay was spread across the inside of the dam wall to help make a better seal.

Toward the end of construction, a torrential rain filled the dam halfway, giving hope that the seal was good. After the wall was completed, the swales were
dug to feed water into the dam. A spillway was also placed in the swales to release excess water. At present, only 10 percent of the swales are completed. When
they are completed, however, they will give the system an additional 170,000-liter
(45,000-US-gallon) capacity before water goes over the spillway.

There was a bit of trouble with the dam. At some point around the halfway
mark, the dam was leaking water. To deal with this, we successfully used bentonite
at a rate of 1 pound per square foot. Construction also ran into some stoppages due
to wet conditions. This interfered with proper compaction and also made operating
the loader impossible, as it would spin on the wet, clayey soil. Unfortunately, the
schedule was set by the contractor's availability, meaning that holding off construction for more optimal conditions was not possible.

A severe drought struck the first summer after construction, and the dam as-
sisted in providing irrigation for the farm through the many dry weeks. There are
plans to complete the swales for the site and to add additional water-harvesting
elements to the farm.

References

Cobb, Fiona. *Structural Engineer's Pocket Book*. Second Edition. Oxford: Elsevier, 2009.

Evett, Steven, Paul Colaizzi, and Terry Howell. "Drip and Evaporation." USDA-ARS, Soil and Water Management Research Unit Conservation & Production Research Laboratory. ksre.k-state.edu/irrigate/oow/p05/Evett.pdf.

Hutchinson, Lawrence. *Ecological Aquaculture: A Sustainable Solution*. East Meon: Permanent Publications, 2005.

Nelson, Stephen A. "Slope Stability, Triggering Events, Mass Movement Hazards." 2013. tulane.edu/~sanelson/Natural_Disasters/slopestability.htm.

Radulovich, Ricardo, Schery Umanzor, Rebeca Mata, and Desiree Elizondo. "Aquatic Agriculture: Cultivating Floating Crops on Lakes." *World Aquaculture Society*, 2015-03 doi: 10.13140/RG.2.1.1265.5521.

Cautions

9

On October 21, 1966, 144 people, including 116 children, were killed in a landslide on the slope of Merthyr Tydfil in Aberfan, Wales. The local coal mine had produced enormous piles of rubble, which were perched above the town. In the morning, shortly after students had arrived at school, a 107,000-cubic-meter (140,000-cubic-foot) section of a debris pile 220 feet in height broke free, liquefied, and slid into the town, burying the school and many other buildings. What set of dynamics triggered this slide, and what can we learn from this tragedy so that we might avoid setting off a similar slide?

With respect to landslides, there is some bad news that is best to get out of the way early. They are not very predictable events, and all soil types have the potential for a landslide. Furthermore, on a geological time scale, every mountain and every hill is coming down eventually. As population increases, more and more people are forced to seek out land that had previously been avoided, including landslide-prone landscapes. Not only are people more at risk from landslides due to increasing habitation of landslide-susceptible areas, they are also more frequently triggering landslides with their activities. Landslides kill 14,000 people worldwide per year on average, with the worst on record being the 1920 landslide in Gansu, China, which caused 200,000 fatalities. In addition to loss of life and limb, property damage costs billions of dollars per year. Deforestation in the hunt for resources and land clearing for agriculture have greatly compounded the risk. Road construction, housing construction, retaining walls, culverts, and other development can also inadvertently set up conditions leading to a landslide. Our goal here will be to avoid adding to this process and to prevent it, if possible.

Like the hydrological cycle, the geology of a site is also a flow system. The geological formations are not a static feature. As with water, the soil—and even the rock itself—moves. In that the rock and soil carry nutrients necessary for life, this movement is a part of the nutrient cycle within an ecosystem. The occurrence of a landslide often exposes vital nutrients such as phosphorus and calcium, which encourage plant growth and are actively sought out by terrestrial animals. A slide will add fertility as the debris flows into an aquatic system or runoff water flows through the debris. While these positive effects exist, they are a problem when the slide itself causes loss of life and property.

The dynamics of slides

We can think of a slide as the outcome of two forces—a driving force and a resisting force—in which the driving force overpowers the resisting force. The driving force is the force of gravity acting on the earth, pulling it downward. Soil rests on top of either more soil or on rock. This underlying layer pushes back against the layer above, preventing a slide. It is the force of friction between rock and soil, or between soil particles, that holds the mass of a hillslope in place. When two opposing forces meet, they create a shear force, where the two forces act in separate directions. When the driving force of gravity overcomes the resisting force—generally with the help of outside factors—a shear failure occurs, and a portion of the soil breaks free and slides along a shear boundary that we can think of as separating the soil into two "camps" in terms of force. The section of hillslope that represented the resisting force stays intact on the hill, and the section of the soil representing the driving force falls away.

This separation of sections on hillslopes due to shear can occur in one of three non-exclusive ways: a rotational slide (AKA a slump), a translational or planar slide, or a flow in which the soil behaves like a liquid. These are non-exclusive because more than one type of slide may occur in a single event. For example, a slump can trigger liquefaction, causing a flow, and might also trigger a planar slide.

In a rotational slide, or slump, an arc-shaped shear boundary fails, causing a rupture in the hillslope (see Figure 9.1). Having broken free of frictional restraint, the earth slides down in a chute from its previous location and settles lower down the hill. At the crown of a slump, you will often find a scarp. The profile of a slump looks like a giant seat built into a hillside. There will be a characteristic divot as well as a bump underneath from the deposit of material ejected from the slide site.

Planar, or translational, slides occur along relatively linear shear boundaries that are roughly parallel to the slope face of the hill. When the driving forces overcome the resisting forces, a slab of material breaks free and slides downhill. These slides are common in shallow soils with a slow-wearing granite parent rock.

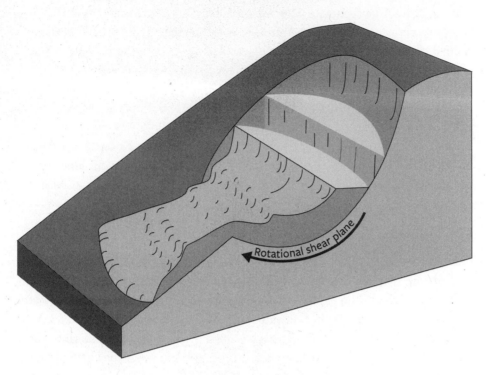

FIGURE 9.1.
A rotational slide,
or slump.

Earth flows occur when the soil moves as a liquid mass down a gradient. One example of this would be landslides in quick clays (AKA Leda clay), which were introduced in Chapter 5, and expanded upon in Chapter 6. Saturated quick clays can undergo liquefaction, in which they rapidly take on a fluid state, leading to free-flowing soil. In the case of liquefaction, flat or nearly flat sites can suffer a landslide. All that is needed is a lower point in the landscape for the soil to flow to, such as a lake or a river, and large portions of land and any accompanying structures can be lost in a muddy liquid flow. While quick clays are the most likely soils in which such liquefaction occurs, all clays can liquefy under the right conditions. In a slump or planar slide, the leading edge of material will often behave like a flow. The leading edge of the deposition zone from a slide will often fan out as though the soil had liquefied.

All three types of slides occur often along boundaries of rock and soil, or one soil-type layer and another. Slides can occur in uniform soils, however. The determining factor is the formation of a shear boundary and some event to trigger a failure along that boundary. The nature of the boundary along the shear plane affects the degree to which it is at risk. For instance, an irregular surface between soil and bedrock will generate more frictional resisting forces against sliding than a smoother surface. Soils have a better cohesive strength than does rock, which gives them more shear strength compared to rock.

Soil creep is another form of earth flow. Creep is a continual movement of earth downhill. Trees growing on these hillsides will have curved trunks as the trees constantly adjust to being offset by the moving earth. Similarly, poles and fence posts will develop a lean as the earth gradually flows downhill. Frost creep will create a rippled washboard-like surface, or mounds that look like small moguls on hillsides, and the freeze-thaw cycle causes displacement of surface soil on a hillslope. A similar rippled pattern known as cat steps—thought to possibly occur as a result of movement when the soil is saturated—occurs in Loess soils.

The role of water

Water-harvesting earthworks always carry an element of risk. Water is one of the major contributing factors to the triggering of landslides. Below saturation, the surface tension of water holds soil particles together more tightly. At saturation, water in the pore spaces of soil puts outward pressure on the soil particles, reducing the cohesive frictional forces between soil particles. If you've ever made a sand castle, you know that you cannot form dry sand. If you wet the sand, you can build walls out of sand that will not collapse, as the water will bind the sand particles together. If a wave should reach the sand castle, the sand will become saturated, and the castle will collapse.

Chapter 3 introduced the concept of the angle of repose, which is the angle at which a soil type will fall to and become stable. A soil wetted to some point below saturation will be more stable than the same soil dry. When it is wetted to the point of saturation, however, the angle of repose will be lower. A pile of soil that is at that soil's dry angle of repose will collapse when saturated with water, resulting in a lower angle of repose. In the case of hillslopes, if there is a boundary of two soil types where one of the soil types becomes saturated, a slide can occur. Water also makes the soil heavier, and additional weight can be a trigger for a landslide. Up until the saturation point, the water will help to hold the soil together; once the saturation point is reached, the soil is not only less stable due to pore pressure, it is also heavier, putting it at greater risk of a slide.

With respect to earthworks, rapid emptying can cause slumping inside a pond or dam. The bank soil is saturated with water, but when it is drained it does not have the support of the open water to hold back the bank. Due to the pore pressure in the soil, the particles are essentially floated, meaning less inter-particle friction (see Figure 9.2). Slumps can then occur in the bank. This can happen in the natural environment too, such as when flood waters rapidly recede. The less steep the slope of the bank, the more stable it will be, and the less prone to such slumping.

Sensitive clays

Canada, Norway, Sweden, Finland, Russia, and Alaska all have regions of quick clay that have been the underlying cause of landslides. These are marine clays that rose above sea level as the land rebounded following the melting of glacial cover after the last ice age. Sodium was the ion that bonded the clay together while in salt water, but gradual leaching by rainwater or groundwater flows left the structure of the clay particles in place while removing the stabilizing salt. Although quick clays can exhibit significant shear strength, when disturbed they can rapidly undergo liquefaction. Furthermore, the collapse of liquefying clay is enough to trigger liquefaction in the neighboring quick clay, setting off a chain reaction that can lead to the collapse of a large area of land. This type of collapse is known as a retrogressive collapse because the scarp formed by the collapse travels backward as more land collapses. The trigger for a 22-hectare (54-acre) slide on relatively flat ground in 2010 in Lyngseidet, Norway, for example, is suspected to have been caused by fill being loaded on the coastline. From the coast, the edge of the slide traveled inland, consuming a greater and greater area, until the slide finally halted.

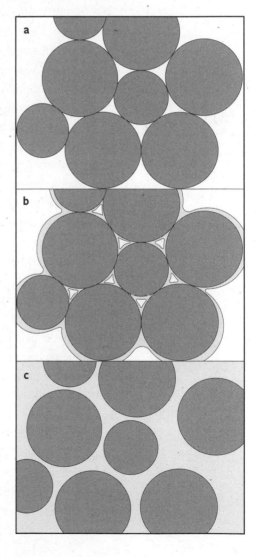

If your region has hard water and a limestone base, calcium and magnesium cations from the rock will stabilize the clay, preventing the formation of quick clay. Conversely, however, if you have soft water, such as is found in areas with a granite parent rock, the sodium and potassium cations that bind quick clay are leached out, making it more sensitive.

FIGURE 9.2. The effects of water on soil. (*a*) Frictional forces between grains hold the soil particles together. (*b*) Frictional forces between particles provide stability. Additionally, the surface tension of water provides an additional cohesive force that makes the soil more stable than when dry. (*c*) The soil is fully saturated. Pore pressure pushes the particles apart, making the soil unstable.

Water-harvesting earthworks are not recommended in sites with quick clay. The further weighting and saturation of quick clays, or the act of construction alone, can trigger a slide. To help stabilize quick clay, potassium chloride can be injected through bore holes at 10.5 kilograms per square meter to stabilize the quick clay on a site to protect against retrogressive slides encroaching on the site. Trees can also be helpful in halting a liquefied earth flow. Should the collapse consume the trees themselves, the flowing trees can jam together, reducing or halting the flow.

Landslide triggers

There are many contributing factors that can contribute to landslide occurrence. The slope of the hillside is a major determinant in risk. The steeper the slope, the greater the chance the driving force can overcome the resisting force, resulting in a slide. For this reason, it is recommended that most earthworks be limited to slopes of under 20°, though the safe operating limit of most machines will be under 15°.

Other factors are the mineralogy and shear strength of soils, ground water flow, hydraulic conductivity of the soils, soil density, pore pressure of the soils, rainfall patterns, aspect of the hillslope, seismic history of the site, site vegetation type, the contributions from plant roots, and the current and past history of land use. The factors directly affecting the soil have been discussed, and their impact on the cohesiveness of a hillslope is readily apparent. The contribution from aspect, however, may not be as self-evident. Aspect is the direction the slope faces with respect to the sun. The more directly the sun falls on a slope, the warmer the soil temperature for that slope. Warmer temperatures mean greater evaporation, meaning drier soils. This, in turn, means that more rainfall is needed to saturate the soil than a slope with less solar gain, giving that slope greater stability, all other factors being equal. Vegetation does have the potential to trigger a landslide, but taken on the whole, the benefits of vegetation with respect to slope stability far outweigh the small chance that vegetation will be the trigger of a slide.

Heavy prolonged rainfall is a major trigger of landslides. Continuous precipitation can saturate soils in addition to weighting them. This increases the driving forces and reduces the resisting forces, increasing the risk of slides. Many regions have careful monitoring of slide-prone areas and issue warnings during periods in which high volumes of rain are falling over a long period.

A site's history is very important with respect to the probability of future landslides. The single greatest predictor of a future slide is a record of a past slide. The history of seismic activity is also important, as earthquakes are often the trigger that sets off many slides. The history of land use is also a critical factor. If a site is cleared of trees, the duration that site has been denuded of trees affects the risk of sliding. The longer the site has been exposed, the greater the risk, in general. The addition of roadways, terraces, and buildings can also increase the risk of slippage. Roads, houses, and terraces can adversely affect site hydrology, increasing risk. Roads and buildings also introduce additional weight loads to a slope.

These triggering factors are what government agencies use to try to estimate the risk of landslides. Currently, the greatest tools in prediction are landslide inventories. These are databases of slide occurrences, used to illuminate a pattern of risk across a region. The greater the density and frequency of landslides in an area, the greater the risk for that area. National efforts and international cooperation in landslide science is very new, but inventories are rapidly being developed to determine slide risks in most nations.

LIDAR (Light Detecting and Ranging, see Chapter 5) is increasingly being used to assess slide risk. By looking at comparative maps, LIDAR can detect when a formerly stable hill is starting to creep. This indicates that shear failure is taking place, and a slide can occur. In addition to early detection of slope movement, an experimental approach for using LIDAR to build landslide inventories has been developed. One indicator of past slides is variation in vegetative growth on a site, particularly trees. Airborne LIDAR is used to image slopes, mapping vegetation. By examining variation of tree height and position, researchers have been able to detect past slides. Slides create a clean slate that must then reestablish with ground cover and then trees. LIDAR imaging helps to discern the areas where this has occurred.

Post-slide treatment

Permaculture often involves restoration of damaged lands. The skill set for this can be applied to sites that have suffered a slide. The first step is to stabilize the soil. A scarp will have formed at the top edge of the side. If possible, cut this edge down to a smooth profile to control erosion and reduce the risk of a future slide. The toe of the deposition zone may also need to be similarly smoothed and graded. Saturation of the site could lead to further instability, so it may be necessary to divert runoff away from the slide to reduce risk of further sliding. Ground cover will need to be established. This might require the addition of soil amendments to assist in plant growth. Compost, kelp meal, or specific mineral amendments may be needed. Mulching or hydromulching the site will help to curb erosion until the ground cover can establish itself. Include nitrogen-fixing plants such as clover or vetch in the ground cover mix. Plant a mix of pioneering trees (which can include nitrogen-fixing trees) as well as long-term, overstory trees. Over broader areas, providing bird perches will help to establish trees on the site. Birds will carry seed in their droppings, which help to provide the site with tree cover. The droppings themselves provide additional nutrient inputs for the soil.

What went wrong at Aberfan?

The Merthyr Vale Colliery sat adjacent the town of Aberfan. The spoils from this coal mining operation were dumped on the hillside above the town, forming very large mounds. The spoil mounds were uncompacted and were left to rest at the angle of repose they fell to as the mounds were formed, reaching a height of approximately 40 meters (131 feet).

The hill upon which the mining spoils were piled had a record of springs before the piles where made, including an artesian spring emanating from the sandstone under the Tip 7 mound, which was to collapse. On the morning of the collapse, workers noticed that Tip 7 had dropped by 3 meters (10 feet). After conferring with the mine office, it was decided that dumping on Tip 7 was to cease. By the time the message got back to the workers at the top, the mound had dropped another 3 meters. Shortly after, the toe of the mound was seen to be moving downhill gradually before the rapid collapse took place. With the rapid slide, the material in the pile liquefied and flowed downhill for 1,950 meters (1.2 miles), destroying Aberfan's Pantglas Junior School and 18 houses before coming to a halt. The leading edge of the deposition zone left a muddy mass 9 meters (30 feet) deep.

The tragic lesson of Aberfan shows the risk of uncompacted material at the angle of repose in saturated conditions. The decision for the placement of the mine tailings was made as a matter of convenience and cost. For our work, we must not only stick to best engineering practices but must also always assume that our project will result in a catastrophic failure. Assuming such a failure forces us to look at the potential consequences should the worst come to pass.

References

Bishop, Alan W. "The stability of tips and spoil heaps." *Quarterly Journal of Engineering Geology and Hydrogeology*. Vol. 6, 1973 doi: 10.1144/GSL.QJEG.1973.006.03.15.

Geological Survey of Norway (NGU). "Quick Clay and Quick Clay Landslides." 2015-02-11. ngu.no/en/topic/quick-clay-and-quick-clay-landslides.

Razak, Khamarrul A., Alexander Bucksch, Michiel Damen, Cees van Westen, Menno Straatsma, and Steven de Jong. *Characterizing Tree Growth Anomaly Induced by Landslides Using LiDAR*, in *Landslide Science and Practice Volume 1: Landslide Inventory and Susceptibility and Hazard Zoning*. Springer-Verlag, 2013.

L'Heureux, Jean-Sébastien, Ariane Locat, Serge Leroueil, Denis Demers, and Jacques Local. *Landslides in Sensitive Clays: From Geosciences to Risk Management*. Dordrecht: Springer, 2014.

Margottini, Claudio, Paolo Canuti, and Kyoji Sassa. *Landslide Science and Practice Volume 1: Landslide Inventory and Susceptibility and Hazard Zoning*. New York: Springer, 2013.

Moum, J., O.I. Sopp, and T. Løken. *Stabilization of Undisturbed Quick Clay by Salt Wells*. Publication Norwegian Geotechnical Institute 81. Norwegian Geotechnical Institute, Oslo, pp 1–8, 1968. trid.trb.org/view.aspx?id=122603.

Nelson, Stephen A. "Slope Stability, Triggering Events, Mass Movement Hazards." 2013. tulane.edu/~sanelson/Natural_Disasters/slopestability.htm.

Pelted, Dave. "Remembering the Aberfan disaster—45 years ago today." 2011-10-21. blogs.agu.org/landslideblog/2011/10/21/remembering-the-aberfan-disaster-45-years -ago-today/.

Rankka, Karin, Yvonne Andersson-Sköld, Carina Hultén, Rolf Larsson, Virginie Leroux, and Torleif Dahlin. *Quick Clay in SWEDEN*. Report 65, Linköping: Swedish Geotech-nical Institute, 2004.

"Report of the Tribunal appointed to inquire into the Disaster at Aberfan on October 21st, 1966." dmm.org.uk/ukreport/553-01.htm.

Walker, Lawrence R., and Aaron B. Shiels. *Landslide Ecology*. Cambridge: Cambridge University Press, 2013.

Appendices

As a part of the planning process, you will need to make some calculations in order to ensure your design is safe and to get an accurate estimate for the cost of your earthworks. The calculations involve basic arithmetic, and for the more complex formulas, online calculators are available. For many people, making calculations may be the least enjoyable part of the earthworks design and implementation process, but it is an important step that should not be avoided.

Calculating Areas and Volumes

To determine the runoff available on your site, you will need to calculate the catchment area. For excavating, you will need to determine the volumes of earth to be moved. While LIDAR and geographic information systems (GIS) such as ArcGIS can automate a lot of the work, these tools are an additional cost that puts them out of reach for many. There is also an initial time investment in learning to use new software systems. If you don't have access to these technologies, you can still make the necessary calculations with pen and paper, or free alternatives in some cases. Additionally, there are online calculators available for all the area formulas given.

The formulas given below have online calculators available at PermacultureEarth works.com.

Area of a triangle

The area of a triangle is usually given as

$A = (b \times h) \div 2$

where A is the area, b is the base of the triangle, and h is the height from the base to the triangle's point above the base.

A more useful formula for the real world is Heron's formula, in which you measure three sides of the triangle and calculate the area from those measurements. This approach is more useful, as it uses the kind of measurements you will actually take in the field. With Heron's formula, the area is given by

$A = \sqrt{s(s-a)(s-b)(s-c)}$

where A is the area; a, b, and c are the lengths of the three sides of the triangle; and s is the perimeter of the triangle divided by 2. That is, $s = (a + b + c) \div 2$.

Area of a rectangle

$A = \text{length} \times \text{width}$

In calculating an area for excavation, you will typically use rectangles and triangles. But what if you are dealing with a curved shape, such as a contour? If the contours are parallel (which is not usually the case, unfortunately), you can take the length of the contour and multiply that by the distance between the contours. If the shape is more complex than that, use the method outlined in "Complex boundaries" below.

Area of a circle

$$A = \varpi \times r^2$$

where A is the area, and r is the radius of the circle.

Complex boundaries

While you can count curved lines as straight when they are parallel, you are not often likely to encounter situations where this is the case. You can break down a complex boundary into rectangles and triangles, but this can quickly become unmanageable with a large and unusual area. In these cases, there are tools to help you. There are a number of online calculators that work with Google Maps. They allow you to map out an area using multiple points. Google Earth Pro (a free tool) also allows you to map out surface areas based on points. These tools make it easy to map out the catchment area above a proposed project. In one respect, this method is more accurate, as the map takes a two-dimensional image of the land. Rain falls more or less straight down (outside of the presence of strong winds), so the map actually represents the surface onto which the rain falls. If you map out a hillside with a tape measure or a measuring wheel, the surface is on a slope (or multiple slopes). Calculating the area of these 3D surfaces will give a greater area than the 2D plane that intercepts the rain. In other words, a mountain occupying a 5-square-kilometer (1235.5-acre) area on a 2D map might well have a 3D surface area of 10 square kilometers (2,471 acres) or more. It will still receive only 5 km² worth of rain, however, so for the purposes of calculating catchment, a 2D map is better.

Volume of a triangular prism

In calculating the volume of earth that needs to be moved for a pond, dam, or swale, you will be dealing with a sloped edge that you can divide into a prism. The volume of a prism is simply the cross-sectional area of the prism times the length of the prism. In other words, it is the area of the triangle multiplied by the length of the prism. The volume can then be given two ways.

$$V = (b \times h \times l) \div 2$$

where V is the volume, b is the base, h is the height, and l is the length.

Volume can also be calculated using Heron's formula.

$$V = l \times \sqrt{(s\,(s-a)(s-b)(s-c))}$$

where l is the length of the prism; a, b, and c are the lengths of the sides of the triangle; and s is the sum of the perimeter of the triangle divided by two s = (a + b + c) ÷ 2.

Volume of a rectangular prism

As with a triangular prism, the volume of a rectangular prism is the area of the rectangle times its height. It is given by

V = length × width × height

Calculating Runoff Volumes

In determining the size of earthworks, you will need to know how much runoff you can expect to occur on the site. Knowing the expected annual runoff as well as runoff volumes in the largest rain events is helpful. Check the average annual rainfall for your area based on the past decade. Also find the record for the single largest rainfall in one day.

We need the rainfall data to find the runoff amount given by the runoff coefficient. This is the volume of runoff divided by the volume of rainfall. Without taking site measurements of rainfall and runoff, you won't know the true runoff amount. You will find it easier to observe the site during rain events and make your best estimate from the following table.

Surface	Runoff Coefficient C
Forested	0.05–0.25
Sandy (bare)	0.20–0.40
Sandy (with crop)	0.10–0.25
Sandy (grass covered, flat to 1.15°)	0.05–0.10
Sandy (grass covered, 1.15° to 4°)	0.10 0.15
Sandy (grass covered, 4°+)	0.15–0.20
Heavy soil (bare)	0.30–0.60
Heavy soil (with crop)	0.20–0.50
Heavy soil (grass covered, flat to 1.15°)	0.13–0.17
Heavy soil (grass covered, 1.15° to 4°)	0.18–0.22
Heavy soil (grass covered, 4°+)	0.25–0.35

Keep in mind that soils can vary greatly across a site, that runoff increases with slope, and that there needs to be a minimum rainfall volume to generate runoff. A light rain is likely to produce little or no runoff. Stronger rain events and rainfall over a longer duration will produce more runoff. Remember, too, that climate affects runoff volumes. Humid regions generate less runoff than drylands.

Runoff is calculated by multiplying the catchment area by the volume of rainfall, and multiplying this again by the runoff coefficient. The formula is

$V_{runoff} = A \times R \times C$

where V_{runoff} is the volume of runoff in liters, A is the catchment area in square meters, R is the rainfall in millimeters, and C is the runoff coefficient.

The easiest calculations will be in metric, and the results can be converted at the end, if you wish. This formula has an online calculator available at www.Permaculture Earthworks.com.

You can calculate both your annual expected runoff and runoff from a single large event. Knowing the runoff for a large single event is helpful in design because it shows you the magnitude of water volume your site will face. You can then scale earthworks appropriately. With climate change, weather patterns are becoming increasingly erratic. It is in your best interest to add an extra 10 percent to the largest rainfall on record. This will help you design in the capacity for extreme events, making your systems safer and more resilient. For the annual average runoff, you can plan for bad times by subtracting 10 percent of the total runoff volumes. The point here is that you should assume drought, rather than relying on either the average or even mean rainfall. Keep in mind, however, that if you are subject to restrictions on total storage sizes, such as in some Australian states, you must stay within the legal limit for permitted storage. You also need to consider the amount of land you are decoupling from the watershed so as to minimize your impact on streamflows.

References

California Environmental Protection Agency State Water Resources Control Board. Runoff Coefficient (C) Fact Sheet (Fact Sheet-5.1.3). waterboards.ca.gov/water _issues/programs/swamp/docs/cwt/guidance/513.pdf.

Ghosh, S.N., and V.R. Desai. *Environmental Hydrology and Hydraulics: Eco-technological Practices for Sustainable Development.* Enfield: Science Publishers, 2006.

Finding Slopes and Heights

You may find yourself needing to use a little bit of trigonometry to determine either the slope of a bank or hillside, or the height of a bank or a hill. There is a mnemonic device that is very helpful in remembering how to use trigonometric functions: the made-up word SohCahToa. A hillside or embankment has the shape of a right-angled triangle. As long as you know the measurement of two sides of the triangle, you can find the slope in degrees. If you know the slope and the measurement of one side, you can find the lengths of the other sides. If the mathematics is confusing, there are online SohCahToa calculators to help you.

The mnemonic SohCahToa describes the relationship between the sine, cosine, and tangent of the slope angle to the three sides of a triangle. The slope angle is the angle between the base of the triangle and the length of the surface of the hillslope. The "o" represents the length of the opposite side of the triangle, which would be the height. The "h" is the hypotenuse, which is the mathematical name for the length of the surface along the hillslope. The "a" is the adjacent side, which is the base of the triangle. The sine S is found by dividing "o" by "h." The cosine C is "a" divided by "h." And the tangent T is "o" divided by "a." The angle of the slope is traditionally represented by the Greek letter theta, which is θ. In other words $\sin\theta = o/h$, $\cos\theta = a/h$, and $\tan\theta = o/a$. The formulas given below have online calculators available at www.PermacultureEarthworks.com.

If you know the slope, and want the height:

If you have either a digital or an analog inclinometer (see Chapter 5), you can find the angle. With the angle and measurement of the surface of the hillslope, you can find how high the hill is with the following:

$$o = \sin\theta \times h$$

For example, if the slope is 11°, and the distance along the hillslope is 100 m, then the height of the hill "a" is

$$o = \sin(11°) \times 100 \text{ m} = 19.08 \text{ m}$$

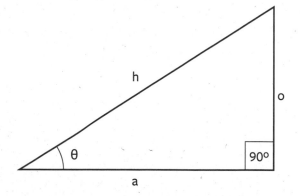

So the height of the hill is 19.08 m. (You will need a calculator, and you will need to make sure it is in degrees, not radians.)

If you want to know the slope:
You will need to know two other sides of the triangle. If, for example, you have the height and the hypotenuse, then

$$\sin\theta = o/h$$

To find the angle, you need the inverse sine (using a calculator):

$$\theta = \sin^{-1}(o/h)$$

For the above example, o = 19.08, h = 100. o/h = 0.1908, and $\sin^{-1}(0.1908) = 11°$.

If you have the adjacent length and the hypotenuse, or the adjacent length and the opposite length, you can use $\theta = \cos^{-1}(a/h)$, or $\theta = \tan^{-1}(o/a)$, respectively.

APPENDIX 4

Swale Spacing

Chapter 7 explained that, for the purposes of groundwater recharge, the distribution for swales on a hillslope should be such that more swales are placed in the upper portion of the hillslope than in the bottom. This is done with a logarithmic distribution of swales. To find the relative position of each swale, we use the expression

$$\log_{(n+2)} S \times D$$

where n is the number of swales in total; S is the swale whose position you are calculating; and D is the distance from the summit to the base of the toeslope, measured along the surface of the hill.

To use the formula, the swale is a assigned a number from the bottom up. The base of the hill at the base of the toeslope is 1; so the first swale up from the bottom is assigned the number 2. If you have 5 swales in total, then n is 5 and the formula for positioning the bottom swale is $\log_7 2 \cdot D$. The second swale from the bottom is given by $\log_7 3 \cdot D$, and so on, with the top swale being $\log_7 6 \cdot D$. You will note that the bottom of the hill is at point $\log_7 1 \cdot D$ (which equals zero), and the summit is $\log_7 7 \cdot D$ (which is 1 multiplied by D), neither of which have swales.

Let's look at an example in which the distance D from the bottom of the hillslope to the top is 100 m. We have five swales.

The bottom swale is at $\log_7 2 \times 100 = 36$. It is located 36 m (rounded off) from the bottom of the hill.

The second swale from the bottom is at $\log_7 3 \times 100 = 56$ m from the bottom.

The third swale is at $\log_7 4 \times 100 = 71$ m from the bottom.

The fourth is at $\log_7 5 \times 100 = 83$ m from the bottom.

The fifth swale, our top one, is at $\log_7 6 \times 100 = 92$ m from the bottom.

Please note that the swale positions are calculated from the bottom up but are dug from the top down.

The size of the swale will depend on the catchment area above the swale and the coefficient of runoff for the site. There can be a lot of variance in runoff, so you will need to observe the site to make your best estimate regarding runoff percentages. The coefficients of runoff for various surfaces are given above in Appendix 2.

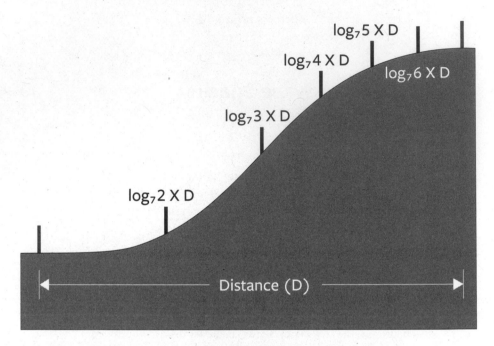

From Appendix 1 above, the catchment area can be treated as a rectangle. Therefore, the catchment area is

A = length of swale × space between swales

The volume of the runoff (Appendix 2) for rainfall volume R and runoff coefficient C is

$V_{runoff} = A \times R \times C$

For example, if we have a 100 m long swale with a 10 m space between that swale and the next uphill swale, then the catchment area is 1,000 m².

If the site has 30 percent runoff, it will have a runoff coefficient C of 0.30. If the maximum recorded rainfall R is 45 mm in one day, then from the formula,

$V_{runoff} = 1000 \times 45 \times 0.30 = 13{,}500$ liters

This means that the swale will need to hold 13,500 liters of water.

We want to find the cross-sectional area of the swale to know how big to make it. There are 1,000 liters in one cubic meter, so this swale holds 13.5 m³. The swale is 100 m long, so if we divide the volume by the length, we will get the cross-sectional area of the swale.

Cross-sectional area = 13.5 ÷ 100 = 0.135 m²

You can play with the dimensions of the swale to fit your needs and preferences, as long as the cross-sectional area has a minimum of 0.135 m² or 1350 cm². Recall that the sides of the swale should have a 1:3 slope (18.43°) to avoid erosion. That being the case, for an absolute minimum volume swale, you could make a simple V-shaped swale that is approximately 128 cm across and 21 cm deep.

Finding the dimensions from the area involves using the trigonometric equations from above and assumes that we are building a V-shaped swale with sides that have a slope of 1:3 (18.43°). So the area of 1350 cm² is a triangular shape that we can divide into two right-angle triangles. Looking at one of the triangles, its area is

$$A_{Triangle} = (b \times h) \div 2 \text{ from Appendix 1.}$$

The total area is two triangles, so

$$A_{total} = b \times h$$

b and h here will give us the dimensions we want for the swale. We know the angle of the sides is 1:3 or 18.43°, and the area is 1350 cm². We can use SohCahToa from Appendix 3 to find first one edge, then the other.

$$\tan\theta = h/b$$

From this, we find the height $h = b \times \tan\theta$. We know $A_{total} = b \times h$, so

$$A_{total} = b \times b \times \tan\theta$$

This gives us

$$b = \sqrt{(A_{total} \div \tan\theta)}$$

In our case,

$$b = \sqrt{(1350 \text{ cm}^2 \div \tan 18.5)} \approx 63.65 \text{ cm} \approx 64 \text{ cm.}$$

Therefore, h is

$$A_{total} \div b \approx 21.21 \text{ cm.}$$

This leaves us with a swale of 128 cm across (64 cm × 2) and 21 cm deep at minimum. In the real world, you will have a hard time getting a bulldozer or excavator to be precise down to the centimeter without spending a lot of time. As a result, you will aim for a swale that is around 130 cm across and 25 to 30 cm deep.

The main assumptions built into the swale spacing formulas are that runoff is uniform across the hill and that the contour lines along the hill are parallel. While not perfect, it serves as the best guide available for swale spacing. Most contour lines are not parallel. If they are too irregular to be taken as parallel, you can use Google Earth Pro or an online

Google Maps area calculator described in Appendix 1 to determine the catchment area. An online calculator that calculates spacing and swale size is available at Permaculture Earthworks.com. Note that the calculator makes no estimation for subsurface through-flow. You may wish to increase the recommended size to account for this and for larger than expected rainfalls.

References

Barnes, Douglas. Swale Calculator: Spacing tool. 2015-09-22. permaculturereflections. com/swale-calculator/.

Terracing

The FAO recommends the formulas below for designing benched terraces with maximum efficiency. To start with, you will need to know the slope of the hill. You will also need to know the width of your benches. If you are going to use tractors on the bench, they will need to be from 3.5 to 8 meters wide (11.5 to 26.25 feet). If you are cultivating by hand, they will be from 2.5 to 5 meters wide (8.2 to 16.4 feet). The final figure you need is the slope of the riser for the terrace. If you are building the terrace with a bulldozer, the riser can be 1:1 (45°). If it is hand built with an earthen riser, it can be 0.75:1 (53.13°). If you build a stone wall, it can be as steep as 0.5:1 (63.43°). For the formulas given below online calculators are available at www.PermacultureEarthworks.com.

The *vertical interval* (V_i) for the terraces is the vertical distance from the base of one riser to the base of another riser. This gives us the horizontal space taken up by one complete terrace bench. The vertical interval is given by

$$V_i = (S \times W_b) \div (100 - S \times U)$$

where S is the slope of the hillside in percent (not degrees), W_b is the width of the bench, and U is the slope of the riser expressed as a ratio.

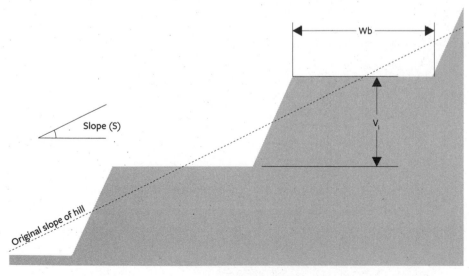

Let's take an example where the hill is 30°. As a percent, the slope is 57.73. [Use the online calculator to convert, or use Percentage = 100 × tan(Degrees).] The bench is to be 2.5 meters wide. We are making a stone wall, so the slope as a ratio is to be 0.5.

$$V_i = (57.73 \times 2.5 \text{ m}) \div (100 - 57.73 \times 0.5)$$
$$= 144.325 \text{ m} \div (100 - 28.865)$$
$$= 2.03 \text{ meters}$$

This vertical interval (V_i) is the same whether you are building level-sloped, reverse-sloped, or outward-sloped terraces.

If you are building a level-sloped bench terrace, the V_i gives you the riser height. Keep in mind that you will likely build a 15-cm bund above V_i at the edge to prevent any water from flowing over the riser and eroding it.

If you are building a reverse-sloped terrace, sloped inward at a 5 percent slope, the riser will be a little higher than the vertical interval V_i. The height difference is the *reverse height*, R_H, and is given by

$$R_H = W_b \times 0.05$$

In our example, this is

$$R_H = 2.5 \text{ m} \times 0.05 = 0.125 \text{ m} = 13 \text{ cm}$$

To get the *height of the riser* (H_r) for a reverse-sloped terrace, simply add the vertical interval (V_i) to the reverse height (R_H):

$$H_r = V_i + R_H$$

For our case, the height of the riser H_r is 2.16 meters. This is approaching the practical limit of the riser height. Keep in mind that the higher the riser, the harder maintenance becomes.

In the case of outward-sloped terraces, the height difference is the *outward height* O_H. Outward-sloped terraces are set at 3 percent (1.72°), so we get

$$O_H = W_b \times 0.03$$

The height of the riser (H_r) is then the reverse height (O_H) subtracted from the vertical interval (V_i).

$$H_r = V_i - O_H$$

References

FAO Corporate Document Repository. "VI. Continuous Types of terraces (Bench Terraces)." *Watershed Management Field Manual: Slope Treatment Measures and Practices.* fao.org/docrep/006/ad083e/AD083e07.htm.

Sheng, Ted C. "Bench Terrace Design Made Simple." 12th ISCO Conference, Beijing 2002. tucson.ars.ag.gov/isco/isco12/VolumeIV/BenchTerraceDesignMadeSimple.pdf.

Costing Earthworks

For earthworks projects there are up to three components to the earthmoving:

excavating and loading + hauling, dumping, and returning + spreading and compacting.

Every project will involve excavation in one form or another, even if it is ripping with a subsoiler or scraping soil to build an embankment or swale with a bulldozer. If you are digging a hole, you will need to excavate the soil and then load that soil to move it to some location. On a large project, it might be loaded into a dump truck. In many cases, loading and hauling are done with a front end loader.

Hauling and dumping are required if you are digging a hole in the ground for a pond or dam. The excavated soil needs to be moved out of the way so that the excavator can move freely. In the case of a dam, the excavated material is used in the construction of the dam wall. In the case of a zone-construction dam, the material will also have to be separated in terms of quality. For pond construction, a location will be needed to place the excavated subsoil. When large piles of either topsoil or subsoil are made, they need management to avoid slides as described in Chapter 9. Bulldozers are needed to push piles down to a safer grade. In dam construction, the soil graded by quality will have to be moved to the wall. In the final stages of the project, the topsoil that was removed needs to be returned to the areas above the waterline. Remember that the machine must make round trips from the soil piles to the excavation site, which costs time, fuel, and money.

Material also needs to be graded and, in the case of dams and ponds needing a seal, compacted. Grading (or spreading the material) requires bulldozer time. Compacting requires time with a roller or other compacting machine. Compaction is done in 15-cm (6-in) lifts, with loose clay typically reducing down to 9.5 cm ($3^{11}/_{16}$ in) on compaction.

Costing will require you to know which steps are needed for your job. When you determine all the steps, you will need to make an estimate regarding the time for each step. There will also be labor costs for your crew and delivery costs for machinery.

Bank, loose, or compacted?

When it comes to moving earth, not all soil is the same. From a machine perspective, it matters if you are moving undisturbed earth, loose soil in a pile that has already been dug, or compacted soil. These three varieties are called "bank," "loose," and "compacted," respectively. When it comes to estimating costs, we measure in cubic meters for each variety.

Bank cubic meters, or BCM, refers to the "bank" or natural, undisturbed soil before any digging, measured in cubic meters. One cubic meter of undisturbed soil is a bank cubic meter.

A loose cubic meter, or LCM, is a cubic meter of soil that has been dug from the ground. The pore space between soil particles increases, and the volume increases. One bank cubic meter will produce more than one loose cubic meter when dug.

A compacted cubic meter, or CCM, is a cubic meter of the soil after it has been mechanically compacted. One compacted cubic meter will be denser than either an LCM or a BCM.

The *swell factor*, or *bulking factor*, is the amount by which soil swells when dug from BCM to LCM. The table below shows the approximate swelling and compressing you can expect for different soil types.

Material	Swell factor BCM to LCM	Volume from LCM to CCM	Volume from BCM to CCM
Clay	43%	63%	90%
Dolomite	50–60%		
Loam	25%	72%	
Sand	11%	86%	90%
Average for most soils	25%		95%

Production rates

To estimate the cost, you will need to know the amount of soil moved per hour by each type of machine for each kind of job it does. This amount will change, depending on whether you are working with BCM or LCM.

The production rate for a machine is the capacity of earth it can move at one time and the number of cycles the machine can perform per hour.

Production per hour = Capacity × Cycles per hour

Both the capacity and the cycles per hour will be affected by both the soil type and whether the soil is BCM, LCM, or CCM. As you can see from the table above, if you neglect to consider whether soil is "bank," "loose," or "compacted," you will be introducing significant error into the estimation. Cycles per hour, or the number of passes a machine can make per hour, are also affected by the skill of the operator. Capacity will also depend on the size of the machine's blade or bucket. To help you in determining a machine's capacity, check the manufacturer's specifications for the machine. Caterpillar, for instance, publishes the *Caterpillar Performance Handbook*. If the manufacturer gives a performance figure in cubic meters, you should assume that they mean LCM. If you are looking at performance for BCM, you will need to convert or time that machine in the given soil conditions to estimate performance. We will look at examples for each machine to give you an idea of how costing is done.

Bulldozer

Let's imagine you are installing a dam. The first thing the you want to do is scrape off and save the topsoil. To do this, you have a track-driven D5N-VPAT Caterpillar bulldozer delivered to your site. Looking at the *Caterpillar Performance Handbook*, we find that the blade has a 2.6 m³ capacity. In other words, the blade can push along a maximum of 2.6 LCM of material.

Let's say that the machine has to clear a 45 m × 45 m patch of clayey material out of the way. For the sake of machinery movement, we need the perimeter of the site clear, too, so we push the material back 55 meters. One cubic meter of clay is 1,620 kg, so the blade will push a mass of 4,212 kg. A quick look at the *Caterpillar Performance Handbook* for drawbar pull vs. ground speed shows that we might expect the dozer to push this load at 4 km/h.

There are 1,000 meters in a kilometer, therefore we can expect one cycle's push time to be

Push time = (55 m / 1000 m/km) × (¼ km/h) × 60 min/hr = 0.83 min.

We must not forget that the machine then needs to back up and push the next row of material. The maximum reverse speed for the D5N is 11.3 km/h, according to the *Handbook*. Considering acceleration and shifting time, let's call that 9 km/h.

Return time = (55 m / 1000 m/km) × (⅑ km/h) × 60 min/hr = 0.36 min.

The *Handbook* tells us that the blade is 3.08 meters wide, so let's round that down to 3 meters. If the topsoil is 10 cm deep, that means clearing 0.10 m × 45 m, which is 13.5 BCM per strip. The blade specification is in LCM, and clay has a swell factor of 43 percent. Therefore 13.5 BCM × 1.43 is 19.3 LCM.

With a blade capacity of 2.6 LCM, you should need 7.4 passes per strip. The blade is 3 meters wide, meaning there are 15 strips. That means, 111 pushes to remove the topsoil. The machine has to back up, so we can add 110 reverse trips to this.

Total time (111 × 0.83 min) + (110 × 0.36 min) = 131.73 min

In other words, you would expect the topsoil cleared in about 2 hours and 12 minutes under perfect conditions. But this is the optimum efficiency. You can expect a skilled operator to be 90 percent efficient, so let's assume it will take

Total time @ 90 percent efficiency = 131.73 min ÷ 0.90 = 146.37 min.

Given no other losses of efficiency, we can estimate the job at about 2 hours, 26 minutes.

The *Handbook* will also give you an estimate for fuel usage in low, medium, and light conditions (with clay being medium). The fuel consumption for a D5N in medium conditions is 11.5 to 16.0 liters per hour. If you are pushing up a gradient, and if the soil is wet, you will use more fuel. You will use between 25.25 to 35.13 liters of diesel.

For simplicity's sake, let's say diesel is $1 per liter. Taking fuel consumption at the middle value of 30 liters, we have $30 for fuel. We also have machine operator costs. Let's say labor is at $25 per hour for 2.5 hours. Now we are at $62.50 in labor, $30 for fuel, and the machine rental business's cost per hour for the machine (or a flat fee by the day). If there is a delivery fee, you can add that on top.

Many machine contractors will quote you a per hour cost, for example $110.00 per hour plus a delivery fee. If that is the case, your job is finished when you calculate the total time needed for the job. If you find it useful for your job, you can calculate BCM per hour, LCM per hour, and cost to move one BCM one meter.

You will have more jobs than topsoil scraping for the dozer. There will be spreading and grading of material as well. You will need to look at the volumes of material for each job the dozer does, calculate the time needed, then sum up the total time for the entire project.

Excavator

There are a lot of factors that influence the efficiency of excavators. The most important factor is going to be an efficient operator who knows how to minimize cycle times for machines without sacrificing load volumes. In our example of a dam, let's say a Caterpillar 312C is delivered to the site. It has a heaped bucket capacity of 0.7 m³. With an excavator bucket, the amount of material that fits in the bucket is not always the same as the rated volume. This difference is known as the *fill factor*. With clay, the fill factor is 80 to 90 percent of the rated volume.

Every excavator will have an optimum depth of cut, which is between 30 percent and 60 percent of the machine's maximum digging depth. Digging outside of this optimum range takes more time. After the bucket digs out a load of material, the machine has to swing to the side to dump the material. The greater the angle of swing, the greater the cycle time will be.

Inside of the optimum depth of cut range, and with an angle of swing from 30° to 60°, the load time for the bucket will be 5 seconds. The swing time, loaded, will be 4 seconds, the dump time will be 2 seconds, and the return swing 3 seconds. The total cycle will be 14 seconds. We will say that the fill factor is 85 percent. The material has a swell factor when disturbed. In our case, it is clay, which expands to 143 percent. To know how many BCM the machine will move, we make the following calculation:

[(3600 sec × Heaped bucket capacity × Fill factor) ÷ Cycle time]
× [50 min ÷ 60 min] × 1 / 1.43

which is

[3600 sec × 0.7 × 0.85 ÷ 14] × [50 min ÷ 60 min] × [1 ÷ 1.43] = 89.2 BCM per hour

Loader

If you have loose material that needs to be moved, the most efficient way to do that is with a loader. If the amount is not too great, you could use a loader mounted on a farm tractor, a skid steer, or the loader on a backhoe. For larger volumes, it saves time and money to have a front end loader on site. Using our example of the dam above, a Caterpillar 950H is driven to the site to handle moving loose materials. Looking at the *Handbook*, we see its bucket has a 2.9 LCM capacity. It has a full-turn static tipping load of 11,073 kg, meaning that if you put this load in the bucket the machine will tip forward. The safe limit for a load on a wheel-driven loader is 50 percent of full-turn, static tipping load, so it can carry 5,536.5 kg safely. With clay at 1,620 kg per cubic meter, this means that the bucket will handle even heaped loads safely. It is a four-speed machine with a four-speed reverse. We are concerned with the forward speeds, which are as follows:

1st gear (forward) 6.9 km/h
2nd gear (forward) 12.7 km/h
3rd gear (forward) 22.3 km/h
4th gear (forward) 37.0 km/h

Let us say that we have sorted piles of material and are dumping high-quality clay into the key of a dam. The storage pile is 30 meters from the dump site, so the loader will stay in first gear at 80 percent of the top speed with a load. (Note that 1 km/h is equivalent to traveling 16.667 m/min.)

(6.9 km/h × 0.80 × 16.667 m/min/km/h) ÷ 60 sec/min = 1.5 m/sec

The return trip, the loader can get into second gear and go a little faster. Let's say 80 percent of second gear.

(12.7 km/h × 0.80 × 16.667 m/min/km/h) ÷ 60 sec/min = 2.8 m/sec

For the 2.9 LCM bucket, the machine has a cycle time of 30 seconds to load at the pile, reverse, then dump in the keyway, and reverse. This means the time will be

Cycle time 30 sec
+ Travel with load (30 m ÷ 1.5 m/sec) 20 sec
+ Return trip (30 m ÷ 2.8 m/sec) 11 sec
Total time: 61 sec

There are always little slowdowns on a job, so let's apply the "50-minute hour," which predicts that you get 50 minutes out of 1 hour of work:

$$[(3600 \text{ sec/hr} \times 2.9 \text{ LCM}) \div 61 \text{ sec/cycle}] \times [50 \text{ min} \div 60 \text{ min}] = 142.6 \text{ LCM per hour}$$

Compactor

When creating a seal, compaction is necessary. In a zoned-construction dam, the key, or core, of the dam is carefully compacted to ensure water retention and a stable dam wall. For our example, we'll say that a Caterpillar CP-563E single-drum, padded compactor was delivered to the site. The drum is 2,130 mm wide. Our keyway is 3,000 mm wide, so more passes will be necessary than would otherwise be the case. We add material in 15-cm lifts. You will need to see how many passes it takes to get the needed compaction. Let's say we would take three passes as our number of passes. The center of the core would therefore get six passes altogether, as the key is 3 meters wide. In other words, each lift added to the key would result in six passes with the compactor. The roller operates at a maximum speed of 11.4 km/h, but this is much too fast for compacting properly. A more effective speed would be 2 km/h. We calculate CCM per hour with the following:

$$\text{CCM/hr} = W \times S \times L \times 10 \div P$$

where W is the width of the machine in meters, S is the speed of the machine in km/h, L is the depth of the lifts added in cm, P is the number of passes, and 10 is the conversion factor to convert meters, km/h, and cm to CCM.

This gives us

$$(2.13 \text{ m} \times 2 \text{ km/h} \times 15 \text{ cm} \times 10) \div 6 = 213 \text{ CCM per hour.}$$

Note that if you are using imperial measurements, your width is in feet, the speed in mph, the lifts are in inches, giving you a conversion factor of 16.3 to get CCY (compacted cubic yards).

With compaction of the keyway, use the compaction rate from LCM to CCM in the table in Appendix 5. Clay compacts to 63 percent of LCM, so each lift will compress from 15 cm down to 9.5 cm. This compression will take place over a number of passes with the compactor (6 in our example). You can get an idea for how much compaction you can expect, based on the height of the wall from the bottom of the cutoff trench. For example, if it is 4 meters, you will compact down 6.35 m worth of LCM (4 m ÷ 0.63), 15 cm at a time. This will be 42 lifts, each of which has 6 passes, making a total of 252 passes with the compactor.

Sum the costs

From your blueprints for the project, you can make an estimate of the total amount of BCM that will need to be excavated, and hence the cost of digging that out of the ground.

For ponds and dams, a certain depth of that will be topsoil that will be scraped off with a bulldozer (or possibly an excavator on smaller jobs). The loose material will need to be hauled, and possibly returned, so calculations for LCM will be needed. There may be compaction costs as well. If topsoil was moved, there will be the cost of spreading and grading that material with a bulldozer. The sum of all the costs for all the steps in implementation will give you an estimate for a project. Be sure to add in inefficiencies like the 50-minute hour. It is also a safe bet to add 10 percent to your final estimate to account for error, unforeseen circumstances, and the human factor.

References
Caterpillar, *Caterpillar Performance Handbook* Edition 36. Peoria: Caterpillar Inc. 2006.

Integrated Publishing. *Advanced Structural Engineering Guide Book.* Table 10-1. Soil Conversion Factors. engineeringtraining.tpub.com/14070/css/Table-10-1-Soil-Conversion-Factors-209.htm.

Purify, Robert L., Clifford J. Schexnayder, and Aviad Shapira. *Construction Planning, Equipment, and Methods.* Seventh Edition. New York: McGraw-Hill, 2006.

The Engineering Toolbox. "Soil and Rock—Bulking Factors." engineeringtoolbox.com /soil-rock-bulking-factor-d_1557.html.

Index

About the Author

DOUGLAS BARNES is a permaculture designer and trainer who specializes in rainwater harvesting earthworks. Trained in Australia by Bill Mollison and Geoff Lawton, he has designed and built earthworks in North America, Japan, and Andra Pradesh, India. Douglas has an interest in complexity theory and systems ecology, educational design and rock climbing. He lives in Tweed, Ontario in a passive solar house he designed and built, and he blogs at permaculturereflections.com.

ABOUT NEW SOCIETY PUBLISHERS

New Society Publishers is an activist, solutions-oriented publisher focused on publishing books for a world of change. Our books offer tips, tools, and insights from leading experts in sustainable building, homesteading, climate change, environment, conscientious commerce, renewable energy, and more—positive solutions for troubled times.

We're proud to hold to the highest environmental and social standards of any publisher in North America. This is why some of our books might cost a little more. We think it's worth it!

- We print all our books in North America, never overseas

- All our books are printed on **100% post-consumer recycled paper**, processed chlorine-free, with low-VOC vegetable-based inks (since 2002)

- Our corporate structure is an innovative employee shareholder agreement, so we're one-third employee-owned (since 2015)

- We're carbon-neutral (since 2006)

- We're certified as a B Corporation (since 2016)

At New Society Publishers, we care deeply about *what* we publish—but also about *how* we do business.

New Society Publishers
ENVIRONMENTAL BENEFITS STATEMENT

For every 5,000 books printed, New Society saves the following resources:[1]

25	Trees
2,291	Pounds of Solid Waste
2,521	Gallons of Water
3,288	Kilowatt Hours of Electricity
4,164	Pounds of Greenhouse Gases
18	Pounds of HAPs, VOCs, and AOX Combined
6	Cubic Yards of Landfill Space

[1]Environmental benefits are calculated based on research done by the Environmental Defense Fund and other members of the Paper Task Force who study the environmental impacts of the paper industry.

MIX
Paper from responsible sources
FSC® C016245